建筑立场系列丛书 No.79

跨越时光的屋顶
When Time Jumps through the Roof

[英] 扎哈·哈迪德建筑师事务所等 | 编
蒋丽 贾子光 | 译

大连理工大学出版社

场馆形态学——材料之间的空间

- 004　场馆形态学——材料之间的空间 _ Gihan Karunaratne
- 010　无人机机场原型 _ The Norman Foster Foundation
- 020　2016墨尔本临时展馆 _ Studio Mumbai
- 030　远景之丘 _ Sou Fujimoto Architects
- 036　香港零碳天地竹亭 _ The Chinese University of Hong Kong, School of Architecture

当时光跨越屋顶

- 042　当时光跨越屋顶 _ Herbert Wright
- 052　德国汉堡易北爱乐厅 _ Herzog & de Meuron
- 072　安特卫普港口之家 _ Zaha Hadid Architects
- 090　麦克唐纳贸易中心
- 100　英国设计博物馆 _ OMA + Allies and Morrison + John Pawson
- 114　马德雷拉古堡 _ Carquero Arquitectura

城市地标建筑新格局

- 122　城市地标建筑新格局 _ Silvio Carta
- 126　马尔默Live大楼 _ Schmidt Hammer Lassen Architects
- 138　鹿特丹Timmerhuis大楼 _ OMA

海运集装箱的改造再利用

- 150　海运集装箱的改造再利用 _ Heidi Saarinen
- 154　城市船舱 _ BIG
- 162　Ccasa青年旅店 _ TAK Architects
- 174　艾杰大学的高新科技园 _ ATÖLYE Labs
- 182　锡安音乐中心 _ Savioz Fabrizzi Architectes
- 188　魔鬼之角 _ Cumulus Studio

- 198　建筑师索引

C3 建筑立场系列丛书 No.75

Pavilion Morphology – Space between the Materials

004 Pavilion Morphology – Space between the Materials _ Gihan Karunaratne

010 Droneport Prototype _ The Norman Foster Foundation

020 MPavilion 2016 _ Studio Mumbai

030 Envision Pavilion _ Sou Fujimoto Architects

036 ZCB Bamboo Pavilion _ The Chinese University of Hong Kong, School of Architecture

When Time Jumps through the Roof

042 When Time Jumps through the Roof _ Herbert Wright

052 Elbphilharmonie, Hamburg _ Herzog & de Meuron

072 Port House in Antwerp _ Zaha Hadid Architects

090 Entrepôt Macdonald

100 The Design Museum _ OMA + Allies and Morrison + John Pawson

114 Matrera Castle _ Carquero Arquitectura

New Configurations for Urban Landmarks

122 New Configurations for Urban Landmarks _ Silvio Carta

126 Malmö Live _ Schmidt Hammer Lassen Architects

138 Timmerhuis, Rotterdam _ OMA

Re-use / Re-model - the Shipping Container

150 Re-use / Re-model – the shipping container _ Heidi Saarinen

154 Urban Rigger _ BIG

162 Ccasa Hostel _ TAK Architects

174 Ege University Technopark _ ATÖLYE Labs

182 Sion Music Center _ Savioz Fabrizzi Architectes

188 Devil's Corner _ Cumulus Studio

198 Index

场馆形态学——材料之间的空间

Pavilion
Space between the Materials
Morph

通常，场馆是一座小型建筑物或纪念馆，并且它可以代表一种文化场景、一个展览空间、聚会场所或一次活动。其关键特征就是对场地占用的临时性，及其随后留下的一种有限的实际而短暂的原型设计烙印。

对建筑师和设计师来说，场馆能够为他们提供一个设计原型的基石或一个实现设计理念和解决方案的设施，随后这些理念和方案可以经扩展应用于不同规模并且更加永久的建筑当中。此外，场馆不仅提供测试新材料或材料组合的机会，而且还会结合这些更加实用的构件来尝试理论性和概念性的设计想法。

对于该类型的临时建筑，许多常见的技术和规划限制条件以及挑战均不适用。在一种新的建筑形态实现之前，临时场馆

Typically, a pavilion is a small structure or monument and it can represent a cultural setting, an exhibition space, gathering place or an event. The defining feature is the temporal nature of its occupation of a site and its consequent legacy of a limited physical and ephemeral archetypal design imprint.
For architects and designers, the pavilion can provide a prototypical stepping stone or an apparatus for ideas and solutions which can later be expanded upon in differently scaled and more permanent buildings. The pavilion further affords the opportunity to not only test new materials or material combinations, but also try out theoretical and conceptual ideas in combination with these more pragmatic elements.
In this type of temporary architecture, many of the usual technical and planning constraints and chal-

无人机机场原型_Droneport Prototype/The Norman Foster Foundation
2016墨尔本临时展馆_MPavilion 2016/Studio Mumbai
远景之丘_Envision Pavilion/Sou Fujimoto Architects
香港零碳天地竹亭_ZCB Bamboo Pavilion/The Chinese University of Hong Kong, School of Architecture
场馆形态学——材料之间的空间_Pavilion Morphology – Space between the Materials/Gihan Karunaratne

类型是理想的试验方式——它能突破和模糊建筑语言之间的界限。场馆设计和施工是体现建筑设计参与的一种手段，而在设计过程中，人际互动是关键因素：一个场地变成一个空间，它在本质上可能体现真正的民主，或至少体现一种更加自由的建筑表现形式。

人道主义建筑是建筑师运用创新方法完成自然或人为灾难过后社区改造和重建任务的建筑，它是一种新兴的设计领域，以协作和跨学科实践为特点。这一建筑类型提供了以场馆建筑进行试验的迫切机会。

lenges do not apply. Before a new form of architecture can be realized, temporary pavilion typologies are ideal ways to experiment – to push and blur the boundaries of architectural language. Pavilion design and construction is a means of engaging in architecture where human interaction is the key: a place becomes a space, which may be truly democratic in its nature, or at least a freer form of expression. Humanitarian architecture, in which architects apply inventive methods to the tasks of rebuilding and restoring communities after a natural or civil disaster, is an emerging field of design typified by collaborative and cross-disciplinary practice. The genre presents urgent opportunities to experiment with the pavilion type.

无人机机场原型，意大利威尼斯
Droneport Prototype, Venice, Italy

　　场馆结构在对城市结构进行调查和干预以及建造或改造公共空间方面，是一种有价值的表现方式。对建筑师和设计师来说，场馆提供了以下机遇：在城市环境中测试、设计新建筑或运营新建筑的方式。

　　大多数标志性场馆建筑都具备说明性及功能性的目标。它们的设计目的是通过创新材料和技术来提升人们对设计力量的认知度，并因此激发人们对未来的思索。

　　诺曼·福斯特基金会设计的无人机机场原型、孟买工作室设计的2016墨尔本临时展馆、香港中文大学建筑学院设计的香港零碳天地竹亭以及藤本壮介建筑师事务所设计的远景之丘，都是建筑试验方面的实践，以及结合了对场地、环境和历史保持敏感度的探索性和诗意化设计，不但大胆起用新的建筑技术，并且在建筑设计中体现了社会公益的想法。

　　诺曼·福斯特基金会的无人机机场担负公开的社会责任并且在其概念方面具有可持续性，并且有信心实现协调经济发展和公共利益的目标。项目应用了首创的无人机技术，为非洲地区，尤其是为缺乏可行道路网络和其他基础设施的国家提供援助。无人机机场是一种模块化建筑系统，在地方社区就可以建造。因为它使用当地未经加固的砖块和夯土瓷砖———一种古老的技术，在施工过程中仅需最小的辅助——该建筑物的建造成本很低，而且必然与周围环境存在一种亲密的关系。"一种简单的建筑形式，数字化世界的一次模拟体验。"可以根据需要在各个地方实现无人机机场项目的扩展，并且它的设计可以抵御地震的冲击。该系统既是理论的试验场，又为检验设计原理提供了机遇，也是一种实用的建筑

The pavilion structure is a valuable medium to investigate and intervene in the urban fabric, and to create or transform public space. For architects and designers a pavilion is an opportunity to test and design new architecture or ways of operating in an urban context.

Most iconic pavilion structures have a declarative as well as a functional purpose. They are designed to increase awareness of the power of design through inventive materials and technologies, and therefore to prompt speculation about what the future might look like.

Droneport Prototype by The Norman Foster Foundation, MPavilion 2016 by Studio Mumbai, ZCB Bamboo Pavilion by The Chinese University of Hong Kong and Envision Pavilion by Sou Fujimoto Architects are exercises in architectural experimentation and speculative, poetic design that combine sensitivity to site, context and history with an adventurist use of building technology and a philanthropic approach to architecture.

The Norman Foster Foundation's Droneport is overtly socially responsible and sustainable in its conception and ambitious in its goal to reconcile economic advancement and communal benefit. The project applies pioneering drone technology to delivering aid to regions of Africa, specifically to countries lacking a viable road network and other basic infrastructure. Droneport is a modular building system which can be constructed by local communities. Because it uses unreinforced local brick and rammed earth tiles – an ancient technique requiring minimal support during construction – the structure can be built at low cost and necessarily has an intimate relationship with its surrounds. "A simple architecture, an analog experience in a digital world." Droneport can be extended at need in a wide variety of localities and is designed

2016墨尔本临时展馆,澳大利亚墨尔本
MPavilion 2016, Melbourne, Australia

设计解决方案。
　　无人机机场这样的建筑往往并不符合传统习惯上的建筑类型：它们有其自身的特殊地方性和生物形态关系。对于诺曼·福斯特基金会来说，无人机机场是未来开发项目的催化剂，并且是具备广泛应用性的、颇具潜力的项目。该项目作为一个模块化结构的拱形建筑在2016年第15届威尼斯国际建筑双年展上展出。正如建筑师所说的，无人机机场是一个"组件包"，只有基础框架和制砖机器被运送到现场。制砖所用的泥土和用于基础的岩石等原材料均在当地采购。拱形砖结构（能够在发展中经济体中被轻易重建的一种建筑语言）可以由社区成员装配，并向当地工人传输重要的施工知识。无人机机场意在创建全面的社区和经济保障节点。它是一种担负社会责任的设计，不仅仅是一个概念，更是发展技能和自主性并且加强社区功能的一种建筑形式。
　　位于墨尔本的2016墨尔本临时展馆由孟买工作室设计，是一次对手工传统和技术的探索和赞颂，它通过工匠们的合作得以实现，并且意在促进可持续性和社区理念。该展馆既是集会场所，也是冥想空间，具备可以过滤日光的轻型骨架。竹子是展馆的主要结构材料，而选择该材料的原因在于其轻盈、灵活和坚固性；它们被捆绑在一起，给人一种清晰的绑扎和工艺感，并能体现将其自身结为一体时所蕴藏的能量。展馆的表面应用了当地的红土。由印度特有植物Karvi制成的遮篷式屋顶的四个角几乎触及地面，形成一个帐篷状的外观。

to withstand seismic shocks. The system is both a theoretical testing ground and occasion for examining design principles and also a practical architectural design solution.
Buildings such as Droneport often do not comply with conventional customary architectural typologies: they have their own idiosyncratic vernacular and biomorphic affiliations. For The Norman Foster Foundation, the Droneport is a catalyst for further development and a project with potentially wide application. It was showcased as a modular structural vault at the 15th International Venice Architecture Biennale in 2016. As the architect put it, the Droneport is a "kit-of-parts". Only the basic formwork and brick-press machinery is delivered to site. Raw materials, such as clay for bricks and boulders for the foundation, are locally sourced. The vaulted brick structure – an architectural language which can be easily recreated across developing economies – can be put together by members of the community, giving local workers important construction knowledge. The Droneport is intended to create holistic communities and anchors of economic security. It is a socially responsible design, not just an idea – an architecture that develops skills and independence and builds up communities.
MPavilion 2016 in Melbourne, designed by Studio Mumbai, is an exploration and celebration of craft traditions and techniques, realized through collaboration with artisans and intended to promote sustainability and ideas of community. The pavilion, both a gathering and contemplative space, has a lightweight, skeletal frame which admits filtered light. Bamboo is the principal structural material, selected for its lightness, agility and robustness; it is tied together to give a legible sense of binding and craftsmanship, and suggest embodied energy holding itself together. Local red orca mud is applied to the surface of the pavilion. The

香港零碳天地竹亭，中国香港九龙湾
ZCB Bamboo Pavilion, Kowloon Bay,
Hong Kong, China

  2016墨尔本临时展馆是一座优雅的建筑，它的外形仿佛是周围植物园的一部分，以一种诗意的形态环绕着周围的景观。从本地采购的青石地板为建筑物营造了一种发冷光的效果。人在馆内走动，既能欣赏到景色，也能感受到建筑的环抱。在场馆的中央，一根带顶的金色柱桩将建筑与马路边沟连接起来，突出当地社区与水之间的紧密联系。在展馆的设计当中暗示了人们对手工制作、收集或搜寻获得的东西的关注。通过这种方式，这座轻型建筑与历史、回忆以及景观产生了联系。它同时也表达了建造者和材料之间的具体关系。

  相比孟买工作室的工艺实践，香港中文大学建筑学院设计的香港零碳天地竹亭向人们展示了他们是如何引进数字制造方式从而使一种濒危的传统——竹棚架建筑工艺恢复生机的。该项目探究了如何将计算机建模工具战略性地融入现有的施工过程，以提供一种发人深思的、有创造性的建筑解决方案。该展馆可以用于宣传那些支持低碳生活和施工策略的活动、展览和演出。

  由藤本壮介建筑师事务所设计的上海远景之丘临时结构是一个简约、透明、漂浮状的脚手架式钢结构，它鼓励参观者在其中攀爬或漫步，去发现场地中不同的视角。这个上海的临时项目与藤本壮介2013年设计的蛇形画廊展馆的设计类型和建筑语言是一样的，都为一种概念性结构矩阵，通过一种透明的景观造型从地面上升起。这两个结构都属于中介空间的组合，或者说是"缘侧"空间。

canopy-style roof made of Indian Karvi almost touches the ground on the building's four corners, imparting a tent-like look.

MPavilion 2016 is an ethereal structure elevated as part of the surrounding botanical gardens – a poetic form framing the surrounding landscape. The locally sourced bluestone floor gives the structure a luminescence. As you move through the building, you see vistas but you also experience enclosure. At the center of the pavilion a covered golden peg connects the structure to the water table, emphasizing the local community's close relationship with water. There is a suggestion, in the pavilion's design, of things being handmade, or collected or foraged. In this way, the lightweight building connects to history and memory, as well as to landscape. It also expresses a tangible relationship between maker and materials.

In contrast to the craft practices of Studio Mumbai, the ZCB Bamboo Pavilion by The Chinese University of Hong Kong demonstrates how an endangered tradition – the craft of bamboo scaffolding construction – can be invigorated through the introduction of digital fabrication. The project investigates how computational modeling tools can be strategically assimilated into existing construction processes to provide a thought-provoking and creative architectural resolution. The pavilion was fabricated to promote events, exhibitions, performances that endorse a low-carbon living and construction strategy.

Shanghai Envision Pavilion by Sou Fujimoto Architects is a minimal, transparent, floating, scaffold-like steel structure which encourages visitors to climb and roam to discover different viewpoints of the site. The Shanghai Project shares a design typology and architectural language with Fujimoto's 2013 Serpentine Gallery Pavilion, a formal yet conceptual matrix of structures which immerges from ground to rise through a

蛇形画廊展馆，2013年，藤本壮介，英国伦敦
Serpentine Gallery Pavilion 2013 by Sou Fujimoto, London, UK

远景之丘，中国上海
Envision Pavilion, Shanghai, China

　　走近看去，这个堆叠、架高的上海临时项目是密集而不透明的；然而，给人的印象却是通过框架的层次感而变化的一个建筑矩阵。透明和半透明的表面以及网格结构共同形成了一块柔和的白色渐变雾云。这个上海的临时项目提供了一个焦点、一种规模指标、一种空间体验以及一系列的景色。它是一个能够反映周围城市环境状态的聚集地，也是一种有内部使用空间的景观。其空间建立起了人及其体验之间以及人和机构之间的一种新关系。它是一个试图理解即时空间的小型建筑物。

　　这些临时场馆代表了体验式建筑类型和空间概念的范例，它可以深化建筑语言的力量，并强化对社会问题和人道主义问题的影响。它们也是设计过程中的载体和工具，为试验性和探索性的工程提供必需的眼界和机遇。此外，临时场馆也起到了人道主义意识形态的孵化园的作用，它可以通过建筑对话和试验性的建筑技术及当地工艺产生优秀的设计。

　　许多此类的场馆都在探索和响应建筑环境，并通过其结构、体量、开口、框架、边界、缺乏固定装置和振动来表达一种对景观的全新解读。它们生动地展现了一种真实而鼓舞人心的建筑形式。通过减少接合处和分解体量的方式，它们在城市中创造了易于通行的、亲密而有趣的空间。这些场馆证明了许多伟大的建筑理念都始于一个微小的开端。

transparent landscape. Both structures are collections of in-between spaces, or "engawa" space. The stacked and elevated Shanghai Project is dense and opaque on approach; however, the impression as a whole is of an architectural matrix that changes through the layering of the frames. The transparent and translucent surfaces, and the grid structure, achieve the soft, white gradation of a mist cloud. The Shanghai Project provides a point of focus, a measure of scale, a spatial experience, and a sequence of views. It is a gathering place from which to reflect on the state of the surrounding urban context, and a landscape of habitable structures. Its spaces establish new relationships between people and their experiences and between people and institutions. It is a small scale structure which attempts to understand instant space. These pavilions represent a sample of experimental architectural typologies and spatial concepts that can deepen the potency of architectural language and its impact on social and humanitarian issues. They are also design vehicles and instruments for the process of design allowing or demanding scope and opportunity for experiential and speculative work. Furthermore, the pavilion is an incubator of an humanitarian ideology. It can produce good design through architectural dialogue and experimental building technology and local craftsmanship.

Many of these pavilions explore and respond to architectural context, and express a new reading of landscape through their structure, mass, apertures, frames, boundaries, lack of fixtures and oscillation. They exemplify an architecture which is authentic and inspiring. By eroding articulation and dissolving the object they yield accessible, intimate and enjoyable rooms within the city. The pavilions prove that many of the biggest ideas in architecture have very small beginnings. Gihan Karunaratne

场馆形态学 Pavilion Morphology

无人机机场原型
The Norman Foster Foundation

一座无人机机场的全尺寸原型建于威尼斯的兵工厂场地。该方案是诺曼·福斯特基金会的首个项目。该项目旨在打造由无人机机场组成的网络，能够向因缺乏道路或其他基础设施而造成交通不便的地区运输医疗用品和其他必需品。基金会的目标是在2030年之前，让非洲的每个小镇和其他新兴经济体都拥有自己的无人机机场。

这个试点项目将于今年晚些时候在卢旺达推行，卢旺达的自然地理条件和社会环境给该项目带来了多重挑战。首批规划的三座建筑将于2020年之前完成，将使网络能够向卢旺达44%的地区输送物资。该项目后续将在卢旺达全国建设40多座无人机机场。卢旺达所在的中央位置也更利于该项目向周边国家（如刚果）拓展，将挽救成千上万人的生命。

货运无人机

货运无人机在缺乏道路设施的地方最实用。正如移动电话不需要电话线一样，货运无人机能够跨越深山、湖泊和不可通航的河流等地

卢旺达人口1178万，2019年之前建造三座建筑，可覆盖520万人口，占其全国人口的44%
Rwanda's population = 11.78 million Covering 5.2 Million people
44% of the country's population with 3 buildings by 2019

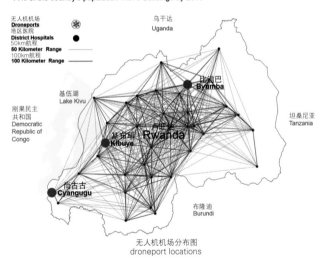

无人机机场分布图
droneport locations

无人机机场的设计和对比
droneport design and comparison

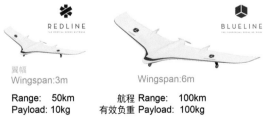

理阻碍，无需建设大型的基础设施。在非洲，仅有三分之一的人口生活在全天候公路方圆2km之内，并且没有高速公路，几乎没有隧道，也没有足够的桥梁，难以抵达生活于非洲大陆偏远地区的人们身边。

专用无人机可以以最低的成本运送血液和救生用品，飞行距离超过100km，为陆路运输提供了一种价格实惠的替代方案。

无人机机场

无人机机场提供了一种新的建筑类型，并且未来有望发展成无处不在的状态，就像加油站一样成为道路运输中随处可见的一项基础设施。它包括保健室、数字化装配车间、邮政和快递室以及电子商务交易中心，因而成为当地社区生活的一部分。

"花小钱办大事"是项目的核心设计理念，而拱形砌体结构实现了最低限度的占地面积，使得当地社区居民可以轻松完成装配工作。还可以将多个拱顶连接在一起形成灵活空间，既能满足特殊要求，也适合无人机技术的发展。

诺曼·福斯特基金会展馆

诺曼·福斯特基金会是一家致力于建筑业及其相关的研究工作实验室的基金会，由它发起建造的威尼斯双年展展馆汇聚了来自欧洲和美国五所大学的教授和学生。

展馆就位于兵工厂的后面，成为通向新开放的公共公园的象征性门户，专门为该项目制作的黏土产品的颜色与其周围历史建筑的颜色相得益彰。

原型结构拱顶的外面两层由这种定制的产品组成，而内层采用了传统的瓷砖。这种特殊产品由稳定的泥土制作而成，这是一种物美价廉而又环保可靠的建筑材料，无需大量使用燃料来实现它的性能。

Droneport Prototype

A full-scale prototype for a droneport has been built on site in the Arsenale in Venice. The droneport prototype is the first

项目名称：Droneport Prototype
项目总监：The Norman Foster Foundation
建筑师：Foster + Partners
合作方：Jonathan Ledgard - Founder of the Pioneering Redline Cargo Drone Network
施工方：Carlos Martin Jiménez - Master Mason, Sixto Cordero, Luisel Zayas, MIT students: Segundo Victor Simba, Luis Alfonso Tiguania Male
施工期限：6 months / 竣工时间：2016
摄影师：©Nigel Young and The Norman Foster Foundation (courtesy of the architect)

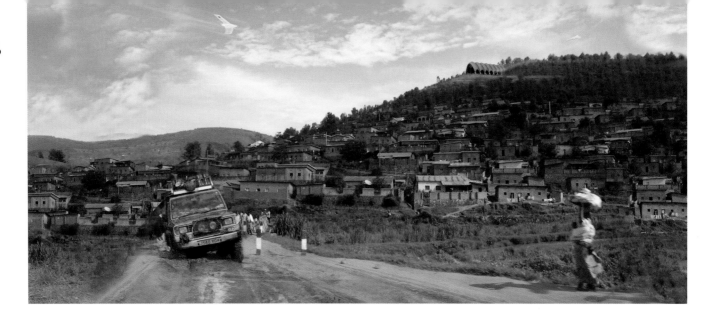

project to be presented by The Norman Foster Foundation. The proposal is to create a network of droneports to deliver medical supplies and other necessities to areas of Africa that are difficult to access due to a lack of roads or other infrastructure. The ambition is that every small town in Africa and in other emerging economies will have its own droneport by 2030.

The pilot project – which will be launched this year – is based in Rwanda, a country whose physical and social geography poses multiple challenges. The initial plan for three buildings, to be completed by 2020, will enable the network to send supplies to 44 per cent of Rwanda. Subsequent phases of the project could see in excess of 40 droneports across Rwanda, and the country's central location could allow easier expansion to neighbouring countries such as Congo, saving many thousands more lives.

Cargo Drones

Cargo drone routes have utility wherever there is a lack of roads. Just as mobile phones dispensed with landlines, cargo drones can transcend geographical barriers such as mountains, lakes, and unnavigable rivers without the need for large-scale physical infrastructure. Just a third of Africans live within two

kilometres of an all-season road, and there are no continental motorways, almost no tunnels, and not enough bridges that can reach people living in far-flung areas of the continent. The specialist drones can carry blood and life-saving supplies over 100 kilometres at minimal cost, providing an affordable alternative that can complement road-based deliveries.

The Droneport

The Droneport offers a new typology for a building, which it is hoped, will grow into a ubiquitous presence, much like petrol stations have become dispersed infrastructure for road traffic. It includes a health clinic, a digital fabrication shop, a post and courier room, and an e-commerce trading hub, allowing it to become part of local community life.

The central idea is to "do more with less" and the vaulted brick structure with a minimal ground footprint, can easily be put together by the local communities. Multiple vaults can also link together to form flexible spaces based on demand and needs of the particular place, and the evolution of drone technology.

The Norman Foster Foundation Pavilion

The creation of a Biennale pavilion was made possible by The Norman Foster Foundation which brought together professors and students from five universities across Europe and America along with a foundation for the building industry and its related research laboratory.

Its location at the end of the Arsenale is symbolic as the gateway to a newly opened public park. The colouration of the earth-based products, which were specifically made for the project, is a careful match with the historic buildings which surround it.

The prototype vault comprises two outer layers of this custom product with an inner layer of traditional tiles. The special product is of stabilised earth – a reliable, affordable and environmentally friendly building material that does not require intensive use of fuel to achieve its performance.

场馆形态学 Pavilion Morphology

2016墨尔本临时展馆
Studio Mumbai

墨尔本临时展馆是由位于澳大利亚墨尔本维多利亚女王花园的内奥米·米尔格隆基金会发起的一个一年一度的建筑设计项目。

在由玻璃摩天大楼组成的都市景观中，2016墨尔本临时展馆摆出了一种被动的姿态，营造了一个发现世界及自我本质的空间。

2016墨尔本临时展馆是一次国际手工建筑设计运动的一部分，建筑师使用7km的竹子、50t的石料和26km的绳子在墨尔本建造了一座独具特色的16.8m²的夏季临时场馆。其设计融合了对传统工艺的兴趣、人类的关联性和可持续性，借鉴了传统技艺、当地建筑技术、材料以及因资源有限而形成的独创性。该展馆使用了专门从印度进口的竹竿，通过5000颗木钉和绳索固定在一起。场馆位于一块始于维多利亚时期的当地采石场的青石板上。屋顶是由利用Karvi的植物枝条编织成的面板建成的，印度的工匠花了4个多月的时间将这些枝条编织在一起。

建筑设计具有通用性，并始终处于一种不断变化的状态，这一点从各种象征和传统中可以体现出来。在未预先设定形象的情况下，展馆的设计经历了不同的阶段，却仍旧保持了对项目功能的敏感度。

墨尔本临时展馆的设计特点在于屋顶的中心设置了一个开口，将自然光、空气和水与大地联系在一起。通过形式与功能的结合，展馆本身的设计以一口金色的水井为中心，象征着水源对于首批定居者的重要性，回归到文明开始的时候，在那时人类并不一定单独存在，而是以群体的形式存在，通过个人对集体的贡献而使集体发展壮大。精心设计的"Tazia"入口塔楼（用于印度仪式中）变成土地与天空、身与心之间的一种交流工具。

2016墨尔本临时展馆不落俗套，利用了印度手工施工方法，通过一系列的模型、材料研究、调查和文献记录对项目精益求精；使参观者沉浸在这些遥远的、相互叠加的景观中，这些景观的存在并不突兀，而是一种基本的元素。该建筑通过这一深植于我们基因当中的如建造一座小屋一样简单的隐性知识启发了社区协作，使社区居民也参与到了展馆的建设当中。通过自力更生的理念（一种获得普通认同的生存之道）的传播，这一建设过程确定了项目基本的道德规范。

该展馆是一个可以让人们会面、交谈、思考和反思的空间。设计的目的不是创建一座新建筑，而是要借着合适材料的选择捕捉空间的精神，表达对周围自然环境的尊重，并与当地工匠协作，以分享设计和施工理念。

本项目以为墨尔本市民建设公共空间的形式来回馈社区民众。

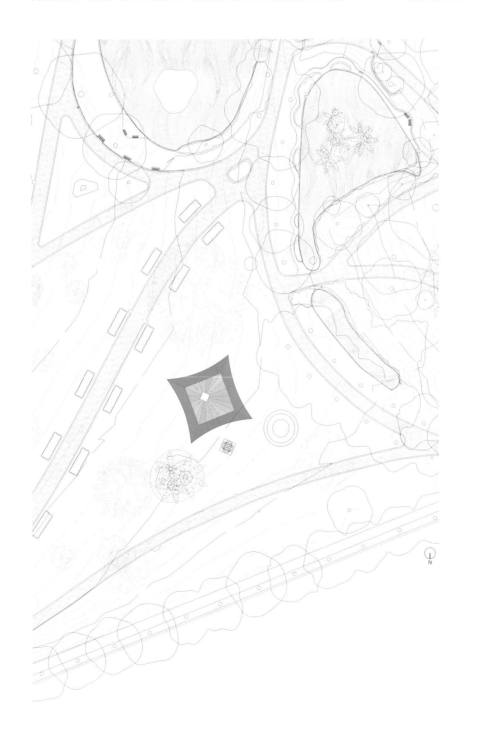

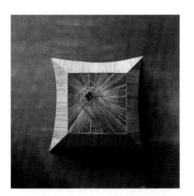

MPavilion 2016

MPavilion is an annual architectural commission by the Naomi Milgrom Foundation situated in Queen Victoria Gardens, Melbourne, Australia.

In a cityscape of glass skyscrapers, the MPavilion 2016 makes a passive gesture; creating a space to discover the essentials of the world – and of oneself.

MPavilion 2016 is part of an international movement in handmade architecture using 7km of bamboo, 50 tons of stone and 26km of rope to create an extraordinary 16.8m² summer pavilion for Melbourne. The design encapsulates its ongoing interest in traditional craftsmanship, human connectedness, and sustainability; drawing from traditional skills, local building techniques, materials and an ingenuity arising from limited resources. The pavilion uses bamboo poles imported from India especially for MPavilion, pegged together with 5,000 wooden pins and lashes of rope. It sits on a bluestone floor sourced from a local Victorian quarry. Slatted panels forming the roof are constructed from Karvi plant sticks woven together by craftspeople in India over four months. Architecture is universal, and is always in a constant state of flux manifested through various symbolism and traditions. With no predetermined image of the pavilion, the design underwent various stages yet sustaining the sensibility of the program.

MPavilion features an opening at the center of the roof that relates light, air and water to earth. Connecting form and

模型 mock-up

项目名称：MPavilion 2016
地点：Melbourne VIC, Australia
建筑师：Studio Mumbai Architects
总建筑师：Bijoy Jain
项目建筑师：Mitul Desai
项目团队：Abdur Rahman, Neelanjana Chitrabanu, Francesco Rosati
客户方项目主管：Robert Buckingham
竹结构顾问：Thumb Impressions Collaborative
Tazia塔楼建造者：Ghulam Mohammed Sheikh's team from Bharuch
施工方：Kane construction
工程：Arup Mumbai, Arup Melbourne
工料测量师：Gardner Group
照明设计：bluebottle
技术咨询：Flot & Jet
客户：Naomi Milgrom Foundation
竣工时间：2016
摄影师：
©John Gollings (courtesy of the architect) - p.20~21, p.23, p.26~27
©Jem Hanbury (courtesy of the architect) - p.28 bottom-left
©Rory Gardiner (courtesy of the architect) - p.25, p.28 bottom-right, p.29

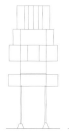
Tazia塔楼立面
tazia elevation

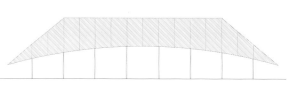
展馆立面 pavilion elevation

Tazia塔楼平面
tazia plan

现场施工 construction on-site

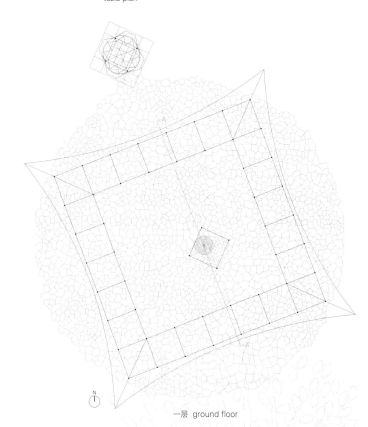

一层 ground floor

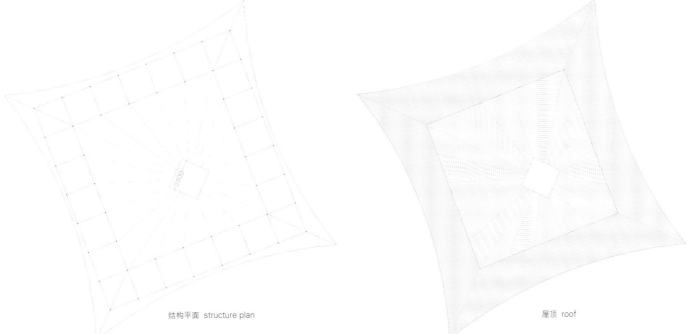

结构平面 structure plan   屋顶 roof

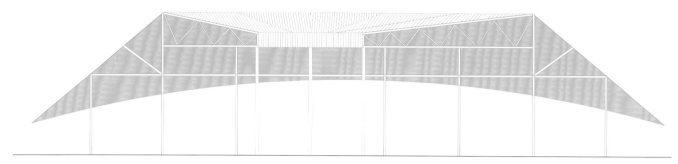

A-A' 剖面图 section A-A'

function, the pavilion centers itself on a golden well symbolising the importance of water to initiate the first settlement, going back to the beginning of civilisation and to the time when man wasn't necessarily an individual but a group that nourished itself through individual contribution. An elaborate "tazia" entrance tower, as used in Indian ceremonies, becomes a tool for communication between the earth and sky, between physical and metaphysical.

Pushing common boundaries, the MPavilion 2016 design employs handcrafted Indian construction methods, refined through a series of models, material studies, research and documentation; allowing the audience to immerse themselves in these distant overlaid landscapes where existence is not reactionary but elemental. The building inspires community collaboration participating in constructing the pavilion through this tacit knowledge as simple as making a hut, which is embedded in our DNA. Such a process identifies the basic and fundamental ethic of the project in the transmission of the idea of self reliance, a way of being which is shared universally.

The pavilion is a space for the people to gather, talk, think and to reflect. The objective was not just to create a new building, but to capture the spirit of the place by choosing the right materials, respecting the surrounding nature and working collaboratively with the local craftspeople to share design and construct ideas.

It's an expression of giving back to the community in the form of public spaces for the people of Melbourne.

远景之丘
Sou Fujimoto Architects

藤本壮介先生为首届"上海种子"项目构建了一个自然与人为元素相融合的建筑环境。从某种角度来看，这种由几何矩阵组成的山脊状结构和矶崎新先生设计的上海喜玛拉雅中心的有机形态得到了近乎完美的契合。游人和架设在半空中的绿植完美融合在一个独一无二的景观之中。藤本壮介在这样一个空间里重新诠释了网格，也重新定义了人与自然环境之间的关系，从而提出了对未来建筑设计的设想。尽管网格是一种体现了人类控制自然界的企图的结构次序，但它也存在于自然之中。因此，这是一种既原始又先进的形式，既具有现代化前期的条件，也具有现代意识。

藤本壮介先生早期发表的论文《原始之未来》(Inax, 2008年)，以及"远景之丘"这样类型的作品，都在呼吁二元化体系的不稳定性 (模糊二分法)，例如原始和现代、自然和文化的对比。尽管我们当下所谓的"未来主义"建筑可能会由以下条件来定义：参数化的造型、智能的外围护结构，或者其他令人叹为观止的技术运用，但是"未来的建筑设计"可能刚好相反，需要返璞归真。

同样，"远景之丘"探索了未来建筑愿景的基本方面。

这座临时建筑的概念始于对未来建筑设计基础的重新构想。这是一种含有两种主要元素的人造有机形式：工业用脚手架和飘浮在空中的树木。

"远景之丘"是一个由白色涂层脚手架构件搭建而成的山脊状三维结构。66棵植物飘浮在这个半透明的景观结构之内，这些白色涂层脚手架构件创造了一个壮观的三维立体的山脊状结构，它23m高，驻立在上海喜玛拉雅中心的正前方。建筑的底层设有画廊、咖啡厅和开放空间，穿梭于这个由玻璃隔墙包裹的精巧建筑之中，同时也面向公众开放。脚手架格子是可以灵活变动的，也容易进出，它创造了一个连续的路径，引导游客穿过这个人造的山形结构。随着人们在空间内自由地闲逛，并与空间产生互动，此次建筑设计成为展示人与自然之间独特关系的平台。

在整个"上海种子"项目期间，在"远景之丘"将举办讲座、联谊会、工作坊、研讨会、演出、展览和电影放映等公共活动。这个开放空间及画廊和安托邦咖啡厅，都将成为这些活动的舞台。它将半透明的景观与飘浮的树木融合在一起，人们在这里得到鼓励去结识新朋友、相互交流，并参与各种活动。

Envision Pavilion

For the Shanghai Project Inaugural, Sou Fujimoto created the Envision Pavilion, a built environment where the natural and the man-made merge. At certain views, the pavilion's geometrical matrix ridge appears to meld with the organic structure of Arata Isozaki's Shanghai Himalayas Center. Through a space where visitors and trees are suspended within a unique landscape, Fujimoto proposes his visions for an architecture of the

courtesy of Shanghai Project

future by reinterpreting the grid and repositioning our relationship to natural surroundings. While the grid is a structural order representing human attempts to master nature, it is also found in nature itself. As so, it is both a primordial and progressive form, a pre-modern condition and modern intention.

Fujimoto's earlier thesis, *Primitive Future* (Inax, 2008), and projects such as the Envision Pavilion, appeal to the destabilization of binaries (to blurring the dichotomy) between primitive and modern, nature and culture. While so-called "futuristic" architecture of our time can be defined by parametric shapes, smart envelopes, or other technological marvels, an "architecture of the future" contrarily may be a primal return. Likewise, the Envision Pavilion explores these fundamentals of a vision for the future.

The concept of the pavilion starts from rethinking the fundamental basics of the future. It is a manmade, organic form composed by two major elements: industrial scaffolding and floating trees.

The Envision Pavilion is a three-dimensional, mountain-like structure made of white-coated scaffolding components. The white coated scaffolding components, setting afloat sixty-six planters within the translucent landscape, create a spectacular (three-dimensional) mountain-like structure that stands 23 meters in front of the Shanghai Himalayas Center. The ground level of this building, where the gallery, café and open spaces are located, extends through the delicate structure enclosed by glazing partitions, and simultaneously opens to the public. The pavilion's scaffolding grid is flexible and easily accessible, creating a continuous pathway to lead visitors across the artificial mountain. As they freely wander and interact with the space, the architecture becomes a platform for unique relationships between human and nature.

For the duration of the Shanghai Project, the Envision Pavilion will act as a site for various public programs including lectures, social gatherings, workshops, seminars, performances, exhibitions and film screenings. The open spaces along with the gallery and Café Entopia will be the stage for these events. They will merge into the translucent landscape with floating trees, where people will be encouraged to encounter and engage diverse interactions and activities.

项目名称：Envision Pavilion / 地点：Shanghai Himalayas Center, No. 869 Yinghua Road, Pudong District, Shanghai, China / 建筑师：Sou Fujimoto Architects Organizing committee: Zhang Lei＿Envision Energy, Dai Zhikang＿Zendai Group, Yongwoo Lee＿Shanghai Himalayas Museum / 艺术总监：Yongwoo Lee＿Shanghai Himalayas Museum, Hans Ulrich Obrist＿Serpentine Gallery / 组织者：Shanghai Himalayas Museum & Shanghai Project / 调查研究：Jenova Chen, Douglas Coupland, Sou Fujimoto, Liam Gillick, Liu Yi, Cildo Meireles, Otobong Nkanga, Xu Bing, Yu Ting, Zhang Haimeng / 总面积：670m² / 竣工时间：2016.9 / 首期临时结构的展览时间：2016.9—2017.7 / 摄影师：©Vincent HECHT (courtesy of the architect) (excepted as noted)

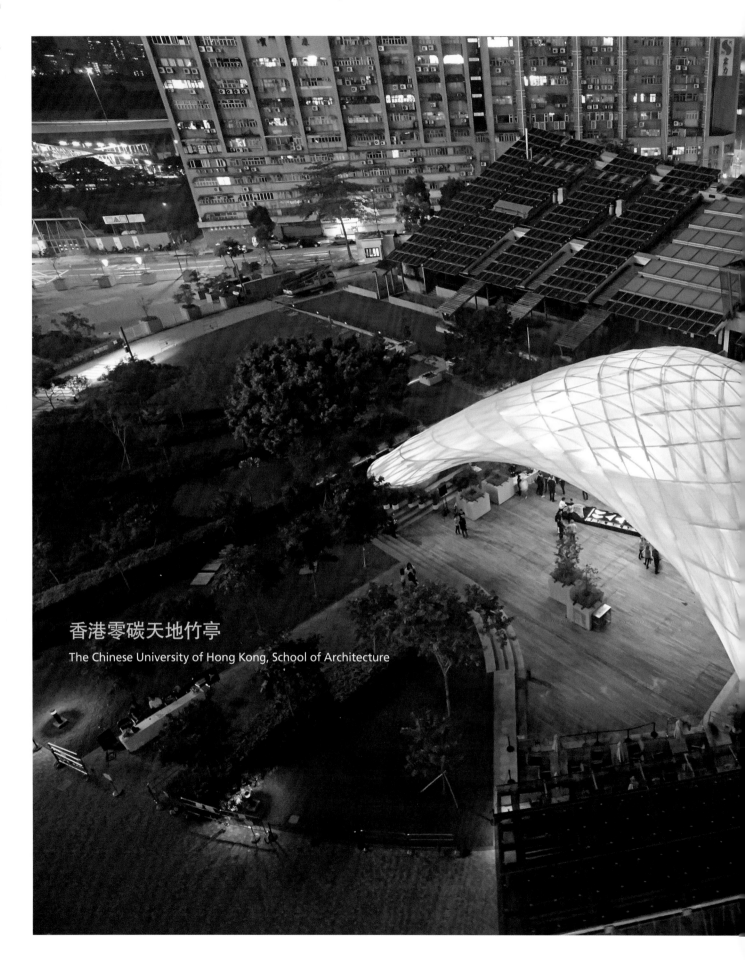

香港零碳天地竹亭
The Chinese University of Hong Kong, School of Architecture

零碳天地竹亭是一个公共活动空间,是2015年夏季在香港九龙湾为建筑工业委员会的零碳建筑(ZCB)而打造的。它是一个四层楼高的主动弯曲大跨度竹质索撑网壳结构,占地面积约350m², 可为200人提供座位。

竹亭由475根大竹竿构成,这些竹竿在现场弯曲以形成结构所需的形状,并借鉴了广东传统的竹制脚手架工艺,用金属线将它们手工捆绑在一起。竹亭是一个巨大的斜肋构架网壳结构,三个角向下折叠组成了三个空心柱,再固定于混凝土地基上,并覆盖上定制的白色弹力布,照明装置安装在三个柱腿的内部,照亮了整个竹亭。

该研究探讨了如何将计算机设计工具战略性地应用于现有的施工方法当中,以便创造出更具吸引力和创新的建筑成果。零碳天地竹亭就展示了这一点,并生动地演示了如何通过数字找形和实时物理仿真工具的引入,挽救和发展香港的竹支架施工这一濒危工艺。

该项目的设计基于香港中文大学建筑学院举办的一次建筑学生设计实习。研究团队随后通过与结构工程师和竹结构顾问的合作,利用数字物理仿真引擎、物理模型制作和大规模的原型制作等手段进一步深化设计,从而实现了最终的外形。

竹子是一种得到广泛使用的环保材料,在亚太地区、非洲和美洲大量种植,成长迅速。它是一种极佳的可再生自然资源,能够吸收二氧化碳,并将其转化为氧气。它坚固、质地轻盈、易于加工和运输。在香港,竹子主要出现在临时剧院、脚手架或宗教节日的结构中,在全球范围内,它通常被用作木材或钢材的替代品,而这些应用都没有利用竹子的独特弯曲特性和强度。相比之下,零碳天地竹亭向人们呈现了一种另类的建筑应用,最大限度地利用了竹子潜在的材料性能。

竹亭的几何形状复杂,其中的每根竹子都有独特的形态、尺寸和结构性质,而且脚手架结构一般都是在没有图纸的情况下利用专业直觉建造的。因此,建造该项目考验的是建筑师对设计的控制能力。为了应对这些不可预见的困难,设计团队开发了一种将精确的数字化设计系统和自然资源相结合的新方法。

零碳天地竹亭向公众和设计界推广创新型和生态建筑设计。它将用于举办倡导低碳生活、施工和开发项目的展览、表演和活动。通过促进大跨度建筑项目的可持续性、轻量化施工方法，寻求广东传统工艺在21世纪的发展方向。

ZCB Bamboo Pavilion

The ZCB Bamboo Pavilion is a public event space built for the Construction Industry Council's Zero Carbon Building (ZCB) in the summer of 2015 in Kowloon Bay, Hong Kong. It is a four-storey-high long-span bending-active bamboo gridshell structure with a footprint of approximately 350m² and a seating capacity of 200 people.

It is built from 475 large bamboo poles that are bent onsite to shape the structure and that are hand-tied together with metal wires using techniques based on Cantonese bamboo scaffolding craftsmanship. The shape is a large diagrid shell structure that is folded down into three hollow columns. These columns rest on three circular concrete footings. A tailor-made white tensile fabric is stretched over the structure and is brightly lit from inside the three legs.

The research investigates how computational design tools can be strategically inserted into existing construction methods to allow for a more engaging and innovative architectural outcome. The ZCB Bamboo Pavilion showcases this and illustrates how the endangered craftsmanship of Bamboo Scaffolding Construction in Hong Kong can be expanded through the introduction of digital form-finding and real-time physics simulation tools.

The project's design is based on an architecture student design internship held at the CUHK School of Architecture. The design was developed further by the research team in collaboration with structural engineers and bamboo consul-

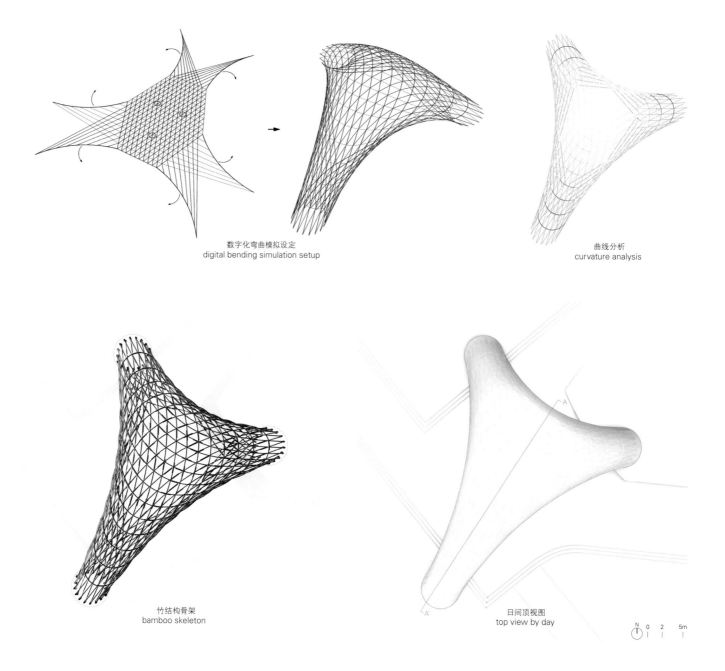

数字化弯曲模拟设定
digital bending simulation setup

曲线分析
curvature analysis

竹结构骨架
bamboo skeleton

日间顶视图
top view by day

项目名称：ZCB Bamboo Pavilion / 地点：8 Sheung Yuet Rd, Kowloon Bay, Kowloon, Hong Kong
建筑师：Kristof Crolla / 项目团队：Dr. Christopher TO _ Excutive director; Yan IP _ Publicity; Margaret KAM _ Technical services / 设计院：The Chinese University of Hong Kong School of Architecture / 总研究者：Kristof Crolla / 合作研究者：Adam Fingrut / 研究助理：IP Tsz Man Vincent, LAU Kin Keung Jason / 授权人：Martin TAM / 顾问：Goman HO, Alfred FONG _ Structural engineering; Vinc MATH _ Bamboo consultant / 结构工程师：George CHUNG / 结构工程师：W.M. Construction Ltd. 总承包商：Sun Hip Scaffolding Eng. Co., Ltd. / 布料结构：Ladden Engineering Ltd. / 照明结构：CONA Technology Co. Ltd. & Brandston Partnership Inc. / 客户：Zero Carbon Building of the Construction Industry Council / 功能：event space / 容积：200 people / 最大跨度：37m / 高度：12.3m (free height of arches: 5.65m) / 重量：6,800kg for bamboo and skin / 竹子用量：475 poles used / 用地面积：425m² / 占地面积：350m² / 施工期间：3.5months / 竣工时间：2015.10 / 摄影师：©Michael LAW (courtesy of the architect) (except as noted)

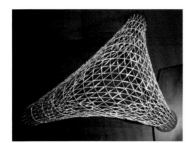

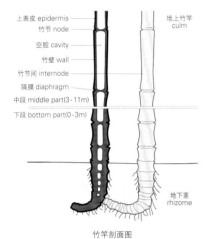

竹竿剖面图
bamboo culm section

tants, using digital physics simulation engines, physical model making, and large-scale prototyping to derive its final form. Bamboo is a widely available, environmentally friendly material that grows abundantly and at remarkably high speeds in the Asia-Pacific region, Africa and the America. It is an excellent renewable natural resource which captures CO_2 and converts it into oxygen. It is strong, light and easy to process and transport. In Hong Kong bamboo mostly appears in temporary theaters, scaffolding, or structures for religious festivals. Globally it is usually applied as a surrogate for wood or steel, rather than in ways that utilize the material's unique bending properties and strength. In contrast, the ZCB Bamboo Pavilion presents an alternative architectural application that maximises these latent material properties.

The pavilion is geometrically complex, bamboo has widely varying geometric, dimensional, and performative properties, and the scaffolding industry does not conventionally use architectural drawings for its intuitive constructions. Building the project therefore challenged the boundaries of the architect's design control. In response, new methods were developed that merged precise digital design systems with inconsistent natural resources in order to deal with these unpredictabilities.

The ZCB Bamboo Pavilion promotes innovative and ecological architectural design to the broader public and design community. It will be used to host exhibitions, performances and events that advocate low carbon living, construction, and development. By promoting sustainable, light-weight building methods for large span architecture project seek ways for traditional Cantonese craftsmanship to evolve for the 21st century.

东立面 east elevation

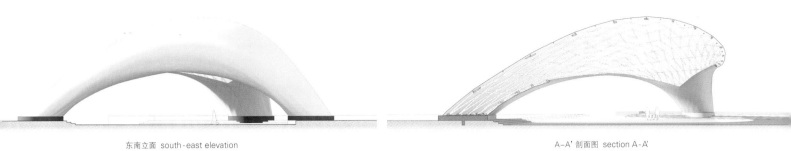

东南立面 south-east elevation　　　　　　　　　　　　　　　　　　A-A' 剖面图 section A-A'

当时光跨越屋顶

When Time through

对具有历史意义（并且通常是作为文物登记在册的）建筑的保护可能与改变其用途的需求和提供一个具有当代品质的内部环境发生冲突。改变建筑用途已经成为建筑实践的主流，但对更多空间的简单需求可能只是要求建造一座新的扩建建筑。然而，扩大建筑占地面积并不一定是建筑师们的选择。对于列管建筑物，旧建筑本身的美感可能是关键的要点——如果横向扩建的话，其具备基本历史价值的周边和外形将会受到影响。

建筑解决方案的目的是保存历史结构，但仅从垂直方向上进行扩建。而通过这种方式设计产生的混合结构

The preservation of historically significant (and usually listed) buildings can clash with the need to change its use and provide a contemporary quality internal environment. Repurposing buildings has become mainstream architectural practice, but the simple need for more space may require a new extension volume. However, extending the footprint is not always an option. In the case of listed buildings, the aesthetics of the old building itself may be the issue - its perimeter and shape have fundamental historic value, which would be compromised by horizontal extension.
The architectural solution is to preserve the historic structure but extend it purely within the vertical.

德国汉堡易北爱乐厅_Elbphilharmonie, Hamburg / Herzog & de Meuron
安特卫普港口之家_Port House in Antwerp / Zaha Hadid Architects
麦克唐纳贸易中心_Entrepôt Macdonald
英国设计博物馆_The Design Museum / OMA + Allies and Morrison + John Pawson
马德雷拉古堡_Matrera Castle / Carquero Arquitectura

当时光跨越屋顶_When Time Jumps through the Roof / Herbert Wright

When Time Jumps through the Roof

为建筑师提供了一次对比当代建筑和旧建筑的机遇。而两者之间的接触面通常位于历史建筑的屋顶轮廓线。

在旧建筑上方扩建新建筑不仅是使已变得多余的建筑恢复生气以及增加新的功能。建筑师可以尽情发挥想象力,采用视觉设计手段创造影响力并吸引人们的关注,并在更广阔的城市景观和城市生活中占据一席之地。本文总结了一些打造这类有历史层次的建筑的设计先例,接着会向大家呈现五个当代的设计案例。

Designing the resulting hybrid structure gives the architect an opportunity to contrast contemporary and old architecture. The interface between the two is usually at the roofline of the historic structure. Building new on old does more than just re-animate what has become redundant and add new functionality. The architect's imagination is free to play visual games that create impact and draw attention, and that can create a presence in the wider urban landscape and the life of the city. This article summarises some of the precedents for creating historically stratified buildings, and then looks at five contemporary examples.

棚户区，南非索韦托，2005年
Shanty town in Soweto, South Africa, 2005

在垂直方向上对建筑进行扩建的历史由来已久，这种扩建方式能够在无须占用新土地的情况下提供新的空间。最初是富有的罗马人为他们的房屋增加额外的楼层，这种扩建方式一直延续到今天，从随着家庭成员增加或生意扩大而需要更多空间的棚户区，到居住者单纯想要获得更多光线和休闲空间的舒适郊区，不一而足。但这种垂直扩建的目的很少是为了彻底打破现有建筑的美感的。

在当代，新的垂直扩建营造了新建筑和古老建筑充满活力的并列状态。追求壮观是当代建筑设计中一个极为常见的设计意图，但如果新建筑位于传统建筑之上，那么就会极大地突出它的视觉震撼性。当代的这种设计趋势可以追溯至1983年，当时，蓝天组开始为位于维也纳Falkestrasse大街转角的一栋19世纪大楼的律师事务所扩建两层办公室。其设计愿景是激进的，建成了一座灵感源于雷电的解构主义建筑。在这个面积为90m²的扩建建于1988年竣工之后，建筑大师查尔斯·詹克斯将其描述为"一只坠毁于屋顶的翼手龙"。而因为寄居于宿主之上，有时它还被描述为"寄生建筑设计"的案例。

继Falkestrasse大街的这个项目之后，建筑师开始在屋顶扩建建筑上自由发挥设计想象。新旧建筑外形、材料和风格之间的对比变成一种有意而为之的设计目标。甚至在颜色对比上也可以动脑筋。例如，2006年建于鹿特丹的Didden村，MVRDV以屋顶之上明亮的蓝色房屋与下方的砖砌建筑之间形成了鲜明的对比。这些是永久建筑物，但是建筑师将许多富有表现力的单层临时构筑物安装在了旧建筑楼上，例如2013年大卫·昆设计的位于伦敦泰晤士河南岸的一处居住设施——怪屋，或2016年艺术家科妮利亚·帕克设计的"过渡对象(Psychobarn)"——一处建于纽约大都会博物馆屋顶花园之上的希区柯克的电影《惊魂记》中房屋的复制品。

在永久建筑物中，许多新旧建筑之间的对话已经从对仓库和工业建筑的改建发展为符合都市生活方式的"阁楼住宅"大楼了。如果建于一栋古老的石砌建筑之上，那么即使是一栋普通的直线型大楼也可以变得激动人心，例如由SHoP设计的位于纽约的Porter住宅(2003年)，或是由KOKO

There is a long history to extend buildings vertically, to provide new space without the need to acquire new land. Wealthy Romans built extra floors to their houses. And the process continues today, from shanty towns where more space is needed as families or businesses grow bigger, to comfortable suburbs where occupants may just want more light and leisure. But such constructions rarely aim to be radical architectural breaks with the aesthetic of the existing building.

In recent times, new vertical extensions have created dynamic juxtapositions of new and legacy architecture. Spectacle is an all-too commonplace intention in contemporary architecture, but the shock of the new is literally underlined if it rests on a traditional structure. The contemporary trend may be traced back to 1983, when Coop Himmelb(l)au started the design of two addition office floors for a law firm occupying a nineteenth-century corner building in Falkestrasse, Vienna. The vision was radical – a deconstructivist structure inspired by a thunder-bolt. After this 90m² extension was completed in 1988, the architectural guru Charles Jencks described it as like "a dead pterodactyl that has crash-landed on the roof". It is sometimes described as an example of "parasite architecture", because it attaches itself to a host.

Following Falkestrasse, architectural imagination was free to play on the roof. Contrasts in form, material and style between old and new became a deliberate objective. Even colour could become the differentiator. For example, at Didden Village in Rotterdam in 2006, MVRDV placed bright blue house forms at rooflevel to maximise the contrast with the brick building beneath. These are permanent buildings, but at this single-story scale, many expressive temporary structures have been mounted on old buildings, such as David Khon's Whimsical Room for London, a habitable installation on the South Bank in 2013, or artist Cornelia Parker's Transitional Object (Psychobarn), a replica of the house in Hitchcock's film *Psycho*, placed in the New York Metropolitan Museum's roofgarden in 2016.

Amongst permanent structures, many dialogues between old and new have developed in the renovation of warehouses and

Falkestrasse大街屋顶改造工程，蓝天组，奥地利维也纳，1988年
Rooftop Remodelling Falkestrasse by Coop Himmelb(l)au, Vienna, Austria, 1988

Didden村，MVRDV，荷兰鹿特丹，2006年
Didden Village by MVRDV, Rotterdam, The Netherlands, 2006

Architects设计的位于爱沙尼亚塔林的Fahle住宅（2006年）。但新的建筑形式显得更加与众不同。最不寻常的仓库的垂直扩建结构并非用于居住用途。Ortner & Ortner设计的位于德国杜伊斯堡的NRW国家档案馆（2013年）是一座76m高的无窗塔楼，其斜屋顶的外形和砖砌外围护结构呼应了下方的旧仓库。而两者之间的对比体现在新扩建建筑陡峭的体量上。Tzannes Associates在位于悉尼啤酒工场的旧锅炉房之上，设计了一家整体采用金属材料的发电厂，该项目竣工于2015年。在所有仓库屋顶扩建项目中规模最大的当属本书中为大家呈现的位于巴黎最长建筑之上的全新城市街区——麦克唐纳贸易中心。

我们可以在博物馆建筑中发现另一种有关形成鲜明对比的垂直扩建建筑的丰富叙述。而旧的基础结构从遗迹到宏伟的公共机构大楼，种类繁多。对于建筑遗迹来说，新旧建筑的接触面不是屋顶，而是由古代建筑遗迹所定义的杂乱线条。这正是由彼得·卒姆托设计的位于德国科隆的科伦巴青砖博物馆（2008年）中出现的情况，新扩建建筑以符合被毁教堂的结构从其上方升起。另一个设计案例是本书中展示的当代项目精选——马德雷拉古堡仅存的两面墙。这些都是受限制的设计作品，但是还有另一个极端，恐怕没有哪一个扩建项目能像丹尼尔·里伯斯金设计的位于德累斯顿的军事历史博物馆（2011年）那样引人注目。它就像一个有棱角的玻璃碎片猛地向上穿透一栋壮观的19世纪新古典主义的兵工厂。

当然，其他的建筑类型也产生了显著的垂直扩建建筑。威尔·奥尔索普设计了一个漂浮于安大略大学艺术与设计学院旧楼之上的、带有像素图案的平板扩建结构（2006年），以扩展公共领域之上的空间。建筑足迹仅在提供支承力的钢支腿接触地面之处发生了微小的改变。由Pir II设计的位于挪威特隆赫姆的Rockheim摇滚乐中心（2010年）以一块四周安装了LED的浮漂板体现了更大规模的设计方案的变化！另一个在其基础结构之上漂浮的扩建结构就是本书随后展示的安特卫普港口之家。

上部扩建结构的建筑负荷所需的额外结构支撑经常要建造在下部古老的历史建筑中。对迄今为止最高的垂直扩建结构来说，情况就是这样。

industrial buildings into urban-lifestyle "loft dwelling" blocks. An unexceptional rectilinear block can become exciting if mounted on an old stone building, such as at The Porter House, New York (2003) by SHoP or Fahle House, Tallinn, Estonia (2006) by KOKO Architects. But the new forms can be far more distinctive. The most unusual vertical warehouse extensions are not residential at all. The NRW State Archives by Ortner & Ortner in Duisburg, Germany (2013) is a 76m-high windowless tower whose pitched-roof form and brick envelope echo the old warehouse beneath it – the contrast is in the sheer massing of the new. Tzannes Associates designed an entire metallic power plant over the Old Boiler House at Brewery Yard, Sydney, completed in 2015. The biggest of all warehouse roof extensions is the creation of an entirely new city quarter on the longest building in Paris, the Entrepôt Macdonald, which the book looks at here.

Another rich narrative of contrasting vertical extensions is to be found in museum buildings. The old base structures range from ruins to grand institutional buildings. With ruins, it is not the roof that is the interface between new and old, but a random line defined by what is left of the ancient structure. This is the case with Peter Zumthor's grey-brick Kolumba Museum in Cologne, Germany (2008), sympathetic in texture to the destroyed church it rises from. Another example is in just two walls of Matrera Castle, included in this book. These are restrained works, but at the other extreme, there is probably no more dramatic extension than Daniel Libeskind's Military History Museum, Dresden (2011), an angular glass shard violently cutting upwards through a grand nineteenth century neo-classical arsenal.

Of course, other typologies have produced notable vertical extensions. Will Alsop floated a horizontal pixel-patterned slab over old buildings in the Ontario College of Art and Design extension (2006), extending its area beyond them over public realm. The footprint can be considered to have changed only marginally – where supporting steel legs touch the ground. The Pir II-designed Rock Music Center Rockheim in Trondheim, Norway (2010) varies the formula on a bigger scale, and with an LED-clad

伦敦怪屋，大卫·昆，Living Architecture and Artangel，英国伦敦，2013年
A Room for London by David Khon, Living Architecture and Artangel, London, Britain, 2013

过渡对象（Psychobarn），科妮利亚·帕克，美国纽约，2016年
Transitional Object(Psychobarn) by Cornelia Parker, New York, USA, 2016

福斯特合伙人事务所设计的赫斯特大厦是一个位于曼哈顿的182m高的玻璃摩天大楼，它采用斜肋构架结构，矗立在场地一处建于20世纪20年代的大厦上方。下方这座旧建筑变成围绕令人震撼的巨大入口中庭的外壳，在其中，摩天大楼的钢框架基础裸露在外。本书随后将介绍的位于德国汉堡的易北爱乐厅项目，是另一个仅保留了新建高层扩建体量下方旧立面的案例，不过，下方的建筑目前绝非虚有其表。

如易北爱乐厅和安特卫普港口之家这类富有表现力的垂直混合建筑，以大胆的姿态融入了城市的天际线。这两栋建筑突出了地方改造，但同时也是对整个城市未来充满信心的表现。但是，像发人深思的马德雷拉古堡所展示的，在旧建筑之上建造新扩建建筑还可以是对过去的陈述。不管怎样，这种混合的垂直建筑物都是一座跨越时光的桥梁。

港口之家位于比利时安特卫普的一个码头区——het Eilandje，该区域在贸易变成以集装箱运输为主之后就荒废了。在许多城市中，时髦的城市再生已经逐渐推进到了这些旧的码头区里，但港口之家的位置与Kattendijk码头是分离的。1922年，一座带有折线形屋顶、实心砖覆层的四层高消防站大楼在这里建造完成。消防站的建筑师Emiel van Averbeke在设计中模仿了安特卫普被烧毁的16世纪著名的Oosterhuis仓库的外观。安特卫普的港务局将其作为办公总部，但该建筑的空间不足以容纳500名员工，所以需要进行扩建。而扎哈·哈迪德建筑师事务所于2008年赢得了该建筑的设计竞赛。

扩建建筑是一个五层高的梯形钢框架体量，长度为111m，从纵向上看去略有不对称，宽度达24m，位于地面之上46m的高度。其建筑面积为6200m²并包括一个可容纳90人的礼堂。建筑外表皮由2000多块三角形玻璃组成，其中70%镶嵌了保温板。在建筑的一端，这些小平面以向外突起的方式安装，呈四面体，借鉴了安特卫普的钻石贸易。新体量由两个混凝土核心提供支撑，其中一个核心建于消防站的庭院中。因为这座水平方向的碎片状扩建建筑引人注目地扩张至旧建筑面向城市一侧的立面上方，所以第二个核心位于旧建筑的外部，并且扩建建筑带有36m的悬挑结构。

floating slab! Another floating extension reaching over its base is Antwerp's Port House, which follows in this survey.

The extra structural support that the building load of upward extensions requires is more often placed within the old historic building underneath. This is the case with the tallest vertical extension yet built. Foster + Partners' Hearst Tower, a 182m-high glass skyscraper in Manhattan in a diagrid structure, rose from the 1920s block on the site. The old building has become a shell enclosing a vast, dramatic entrance atrium in which the skyscraper's steel frame base is exposed. The Elbphilharmonie in Hamburg, which follows here, is another example where only the old facade remains beneath a new high-rise volume, although the lower building is now far from being a void.

Expressive vertical hybrid buildings, such as the Elbphilharmonie and Port House, make bold statements on the urban skyline. Those two buildings highlight local regeneration but are statements of confidence about the future of the whole city. But, as the contemplative Castillo de Matrea demonstrates, building new on old can be statements about the past. Either way, the hybrid vertical building is a bridge across time.

The Port House is situated in het Eilandje, a port area of Antwerp, Belgium, which was abandoned as trade became containerised. As with many cities, hip urban regeneration is advancing into these old docklands, but the Port House stands isolated at Kattendijk Dock. In 1922, a four-storey solid brick-clad fire station with a mansard roof was completed there. The architect Emiel van Averbeke mimicked the appearance of Antwerp's great sixteenth-century Oosterhuis warehouse, which had burnt down. Antwerp's Port Authority chose it as its headquarters, but it was not big enough for their 500 staff – an extension was needed. Zaha Hadid Architects won the architectural competition in 2008.

The extension is a steel-framed trapezoid-like five-storey volume which is slightly asymmetric along its 111m length, up to 24m wide and reaching up to 46m above the ground. It has a floor area of 6,200m² and includes a 90-capacity auditorium. The skin

Porter住宅，SHoP，美国纽约，2003年
The Porter House by SHoP, New York, USA, 2003

Fahle住宅，KOKO Architects，爱沙尼亚塔林，2006年
Fahle House by KOKO Architects, Tallinn, Estonia, 2006

一条开放设计的桥梁连接位于扩建建筑下方的核心，高出地面21m并与旧建筑的屋顶齐平。庭院变成一个中庭，并安装了观光电梯，经由桥梁通向扩建建筑，并且有四根支撑钢柱以不同的角度斜插进中庭。而旧消防站中庭附近6800m²的建筑面积已经过修复。100个80m深的钻孔为港口之家提供地热能源。

安特卫普港口之家的外形就像是某部科幻电影中的形象，而事实上，如果由中庭仰望扩建建筑，它就像是一艘漂浮于人们头顶上的外星宇宙飞船！然而新旧之间的极大反差和充满生机的设计绝对是无与伦比的，在城市的另一边也能一眼就看到该建筑，这标志着安特卫普及其码头已经开始迅速发展。

与其他垂直混合建筑不同，英国设计博物馆原有的屋顶是保留元素，而下方的建筑为新扩建建筑。

1962年，创建于伦敦肯辛顿的英联邦学院，是一栋由罗伯特·马修设计的引人注目的现代主义建筑，该设计师先前负责伦敦皇家节日音乐厅的设计，并且是RMJM建筑事务所的创始人。他以一座底层之上带有不透明蓝色幕墙的正方形建筑将面积为6000m²的博物馆收入其中。可以从位于宽阔的连续阳台的两个展览楼层的曲形边缘向下看到另一个底层的展览层。中央空地同样展现了底面最壮观的特征———一个马鞍形的双曲抛物线包铜屋顶覆层，它由室外两个倾斜的混凝土扶壁提供支承力。屋顶有一个巨大的悬浮的菱形混凝土结构向上倾斜，呈双曲线弯曲并由两个较短的内部扶壁提供支承力。

英联邦学院于2002年关闭。OMA在2008年赢得了将该场地总体规划为住宅区的设计竞赛，该项目将为马修设计的这座标志性建筑的改建提供经费。英国设计博物馆决定搬入该场地。但旧建筑的能源绩效较差，其楼板强度也不足以承受沉重的展品，并且博物馆需要更多的空间。因此OMA提议更换建筑，但保留屋顶和扶壁。新建筑将与旧建筑具有相同的外形。

consists of over 2,000 glass triangles, 70 per cent of them panelled with insulation. At one end, these facets are installed outwards in tetrahedrons, referencing Antwerp's diamond trade. The new volume is supported by two concrete cores, one emerging through fire station's courtyard. Because this horizontal shard dramatically extends beyond the old building's city-facing facade, the second core is situated outside, and there is a 36m cantilever beyond it. An open bridge connects the cores beneath the extension, 21m above the ground and level with the old roof. The courtyard has become an atrium from which panoramic lifts rise through with bridge to the extension, and four steel support columns cut through it at different angles. A further 6,800m² of floorspace has been restored around it, in the old fire station. One hundred 80m-deep bore-holes feed the Port House with geothermal energy.

The Port House looks like something out of a science fiction movie, and indeed, looking up at the extension from the atrium, it feels like an alien spaceship that has floated above you! But the extreme contrast between old and new and the sheer dynamism of the design are simply without equal. Visible across the city, it says that Antwerp and its docklands have taken off.

Unlike other vertical hybrid buildings, the Design Museum's legacy roof is the preserved element, while the building underneath is new.

In 1962, the Commonwealth Institute opened in Kensington, London in a striking modernist building designed by Robert Matthew, previously responsible for London's Royal Festival Hall and the founder of practice RMJM. He housed a 6,000m² museum in a square building with opaque blue screen walls above the ground floor. Two exhibition levels on wide, continuous balconies looked down from curving edges to another ground level exhibition floor. The central void also exposed the underside of the most spectacular feature, a saddle-shaped hyperbolic paraboloid copper-clad roof clad, supported by two inclined exterior concrete buttresses. The roof curves up from a great suspended concrete diamond, also hyperbolically curved and supported by

NRW国家档案馆，Ortner & Ortner，德国杜伊斯堡，2013年
NRW State Archive by Ortner & Ortner, Duisburg, Germany, 2013

啤酒工场的旧锅炉房，Tzannes Associates，澳大利亚悉尼，2015年
Old Boiler House at Brewery Yard by Tzannes Associates, Sydney, Australia, 2015

2010年，极简主义建筑师约翰·波森赢得了在OMA的规划大纲内设计新博物馆的竞赛。用钢结构将原始结构架高至地面之上20m，从而实现新建筑的建造。奥雅纳工程顾问公司负责结构工程部分，包括一个围绕首层的新的钢桁架，连接马修的室内扶壁。新建筑体量楼层面积为10000m²，包括办公室、图书馆和学习室以及一个方形平面的中庭，随着高度的提升，中庭变得更加宽阔，使人们可以看到屋顶。烧结玻璃幕墙与英联邦学院旧楼一样有着蓝色的外部竖框网格，目前为两个立面提供透明度。中庭的一大特色是有一个宽阔的楼梯，中间设计有用作座椅的双层踏板中央分隔带，而中庭的用色很简单，包括木质墙壁和水磨石地面。波森额外设计了一个较小的上空空间，将一个室内扶壁显露出来。地下二层有一个双层高的画廊。

英国设计博物馆的室内空间设计比英联邦学院的更加合理、布置得更加柔和，并且采用的材料不那么原始，但给人的感觉却更加温暖。关键在于无论是室内还是建筑外部，建筑师都没有摒弃对连绵曲折的大屋顶的设计。

位于Villamartín镇（临近西班牙加的斯）的马德雷拉古堡由摩尔人建造于9世纪，后来已是残垣断瓦，所剩无几。遗迹中幸存下来的最大部分为Parjate塔楼的两面支离破碎、杂草丛生的石墙。其中一侧仍留有画着一艘赭色小船的壁画。这一脆弱的结构位于一座小山顶，它是从周围村庄都能看得到的地标性建筑，具有重要的地方文化意义，但这座建筑存在着进一步崩塌的危险。Carquero Arquitectura，塞维利亚的一家年轻的具备修复经验的建筑工作室于2011年接受委托拯救这栋历史建筑。

该项目不仅涉及加固和修复。Carquero工作室选择以新建的光滑墙壁从遗迹上垂直延伸而出，达到塔楼先前的高度。它们与原来的石墙产生了明显的对比，而且新石墙的石灰泥材料是根据从遗址中提取的样本制作的。因为重建与先前石头纹理一样的仿品建筑是受到法律禁止的。所

two shorter internal buttresses.

The Commonwealth Institute closed in 2002. OMA won the 2008 competition to masterplan the site with residential blocks which would finance the repurposing of Matthew's iconic building. The Design Museum committed to moving to the site. But the old building had poor energy performance, its floors were not strong enough for heavy exhibits, and more space was required. OMA proposed replacing the building but retaining the roof and buttresses. The new building would have the same shape as the old one.

The minimalist architect John Pawson won the 2010 competition to design the new museum within OMA's framework plan. It was built as the original structure that was held 20m above ground by steelwork. The structural engineering by Arup included a new steel truss around the first floor connecting Mathew's internal buttresses. The new volume has 10,000m² of floor, including offices, library and learning suite, and a square-plan atrium which widens with height to reveal the roof. A fritted glass curtain wall, with the same external blue and mullion grid as the Commonwealth Institute's, now offers transparency on two facades. The atrium features a wide staircase with a central double-step central strip for seating, and the restrained palette includes wooden walls and a terrazzo ground floor. Pawson has created a second, smaller void to expose an internal buttress. The lower of two basement levels houses a double-height gallery.

The Design Museum's internal spaces are more rational and muted than the Commonwealth Institute's, and the materiality less raw, but the impression it creates is warmer. Crucially, the sweeping drama of the great roof, inside and out, is not lost.

Little remains of the Matrera fortress at Villamartín, near Cadiz, Spain, built in the ninth-century by the Moors. The largest surviving elements of the ruin were two fragmented, overgrown stone walls of the Parjate tower. One side still carried an ochre fresco of a boat. Situated on a hilltop, this fragile structure was a landmark visible from the surrounding countryside, with great local

科伦巴青砖博物馆,彼得·卒姆托,德国科隆,2008年
Kolumba Museum by Peter Zumthor, Köln, Germany, 2008

军事历史博物馆,丹尼尔·里伯斯金,德国德累斯顿,2011年
Military History Museum by Studio Libeskind, Dresden, Germany, 2011

以,最终形成的效果就好似塔楼遗迹融入了一个极简主义的当代构筑物当中。

这一历时五年的项目保护了历史遗迹,并根据建筑的历史情况恢复了它原有的高度,但当项目竣工后遭到了严厉的指责。但是无论如何,这项工程在2016年获得了好几项国际大奖。

麦克唐纳贸易中心项目是城市再生进程中一次重大的合作实践,它建造于一座非凡的预先存在的巨型建筑之上,该建筑堪称巴黎最长的建筑,它是一个三层的现代主义混凝土仓库,沿麦克唐纳大道绵延617m,位于铁路线路和环城大道公路之间。它由马塞尔·福雷斯特设计,占地面积142000m²,位于一排长4m、高6m的混凝土格栅带后方,其下方为卸货区。建筑于1970年完工。

其地理位置使它加入了"缩小旧城和郊区社会住房之间的经济差距"这一城市计划。雷姆·库哈斯及其荷兰建筑师同事弗洛里斯·阿克梅德构思出一个2007年度总体规划,对这栋建筑进行重建,保留建筑原有立面,但也在其上部建造高楼大厦。由阿克梅德的工作室FAA和比利时Xaveer De Geyter建筑师事务所(XDGA)共同完成的修订规划将场地分成不同的部分,供不同的建筑师在旧建筑内部和上部进行设计施工,创建了一座"水平的城市",在原有构筑物之上扩建了多达六层的空间。

15个大多数来自巴黎的建筑团队同步开展此次扩建项目,一条新的有轨电车线路和一座新火车站在施工过程中对外开放。有轨电车从麦克唐纳贸易中心标志性的混凝土格栅带中间的下方穿过。在南侧有一个平行的开发带,包含零售店铺,主要通过巨大的地下停车场上方的狭长花园与北侧的混合构筑物分离开来,但两端的建筑还是使这个165000m²的开发项目在整体上保持了一致性。在场地的西半部,一半是社会住宅,一半是私人住宅,而东半部包括面积为58500m²的办公和商业空间。而1125个住宅单元的总面积为74300m²。

cultural significance, but it was in danger of further collapse. Carquero Arquitectura, a young Seville-based practice with restoration experience, took on the commission to save it in 2011.

The project is more than consolidation and restoration. Carquero extruded the ruin vertically with new smooth walls that reach the height the tower once had. They clearly contrast with the old stones, yet are made with a lime plaster based on samples from the ruins. Recreating the old stone pattern in a copy was forbidden by law. The effect is that as if the ruin is melting into a minimalist contemporary structure.

The five-year project safeguards the ruins and marks its lost historic height without falsifying its history, but it was heavily criticised when completed. Nevertheless, the work won several international prizes in 2016.

The Entrepôt Macdonald is a radical collaborative exercise in urban regeneration, and builds on an extraordinary pre-existing megastructure – the longest building in Paris, a three-storey modernist concrete warehouse stretching 617m along the Boulevard Macdonald between railway lines and the Boulevard Périphérique highway. Designed by Marcel Forest, it enclosed 142,000 m² behind a continuous strip of 4m long, 6m high concrete grills above its loading bays. It was completed in 1970.

Its location made it a candidate for city initiatives to bridge the economic gap between the old city and the social housing of the suburbs. Rem Koolhaas and fellow Dutch architect Floris Alkemade produced a 2007 masterplan to regenerate the building, preserving its facade but with highrises over it. A revised plan by Alkemade's practice FAA and Belgium's Xaveer De Geyter Architects (XDGA) divided the site into sections for different architects to build within and above the old building and create a "horizontal city", up to six storeys above the old structure.

A new tramline and railway station opened as the side-by-side projects of fifteen mainly Paris-based architectural teams were under construction. The tram passes under the iconic concrete grill strip half-way along the Entrepôt Macdonald. A parallel strip

夏普中心，安大略大学艺术与设计学院，aLL Design，加拿大多伦多，2006年
The Sharp Center, Ontario College of Art and Design by aLL Design, Toronto, Canada, 2006

Rockheim摇滚乐中心，Pir II，挪威特隆赫姆，2010年
Rock Music Center Rockheim by Pir II, Trondheim, Norway, 2010

麦克唐纳贸易中心拥有各种不同的建筑风格，从色彩斑斓的后现代风格到平淡无奇的玻璃钢当代现代主义风格，不一而足。而最引人注目的建筑包括由日本建筑师隈研吾设计的大学和体育运动中心，它采用了在架高的广场之上安装胶合板屋顶遮篷的方式，以及由Odile Decq工作室设计的位于初创企业孵化园CARGO、带有圆形窗户的相互堆叠的黑色盒状建筑。从侧面看最狭窄的是由l'AUC设计的学生宿舍，其场地仅有8m宽，并挑空于仓库之上，不过却横跨了从北到南的整个跨度。由Christian de Portzamparc设计的大型住宅区有红色、蓝色、铝质表皮和之字形好几种外观的立面。Hondalette Laporte设计的住宅区上部有一些类似独栋房屋的体量，与MVRDV设计的Didden村相似，但未采用任何色彩。

麦克唐纳贸易中心是打破单独一座狭长大厦单调设计的一次大胆尝试，但它仍能为人们留下一种霸气的感觉。尽管存在多样性，连续而笔直的传统仓库建筑仍能保留其最不寻常的元素。

建于德国汉堡旧码头区的最后一个主要仓库是Kaispeicher A仓库，由维尔纳·卡尔摩根设计，竣工于1963年。该建筑高37m并沿易北河绵延126m，但因采用了梯形的平面设计方案，建筑最短的一侧仅有22m宽。普通的砖砌立面上点缀着小窗户，而全高的垂直卸货区则位于建筑两侧。汉堡市计划对该建筑进行重新开发，但当地的一位开发商经过与瑞士建筑师事务所赫尔佐格&德梅隆的接洽，决定保留这座与众不同的仓库，而探求在其上建造音乐厅的想法。汉堡市最终同意了这个提议，从而也就成就了易北爱乐厅项目。

建筑师设计了一个新的体量，覆层采用预制玻璃面板，并与下方仓库的平面设计保持完全一致。它包括两个音乐厅、一家酒店和豪华公寓，并且建筑顶部设计了波浪形屋顶，其中一端距离街面最高处达到102m。在新扩建建筑的下方，也就是Kaispeicher A仓库的屋顶楼层是一个名为"广场"的公共空间，四周设计了一条开放的人行漫步道，围绕着易北爱乐厅内部的玻璃墙入口门厅。

该建筑当属德国最雄心勃勃的21世纪建筑之一，但自2007年施工开始后，其施工成本不断增加，最后比最初预计成本的七倍还多。仓库的内

of development on the south side, including retail, is separated mainly by narrow gardens over a vast underground car park from the north-side hybrid structure, but end buildings unify the whole 165,000 m² development. On the western half, the developments are an equal mix of social and private housing, while the eastern half includes 58,500m² of office and commercial space. The 1,125 residential units cover 74,300m².

The Entrepôt Macdonald's different components range in styles from colourful post-modern to bland steel-and-glass contemporary modern. The most notable elements include the college and sports center by Japanese architect Kengo Kuma, with plywood roof canopies over a raised plaza, and the stacked black boxes with circular windows at CARGO, a start-up business incubator by Studio Odile Decq. The narrowest section, student accommodation by l'AUC, is just eight meters wide, and cantilevers over the warehouse, but spans completely from north to south. A large residential section by Christian de Portzamparc varies its facades with red and blue colours and aluminium, and zig-zag facades. Hondalette Laporte's residential block has house-like volumes on top, similar to MVRDV's Didden Village but without color.

Entrepôt Macdonald is a bold attempt to break the monotony of a single long block, but it still feels overbearing. Despite the variety, the continuous linearity of the warehouse heritage block remains its most unusual element.

Hamburg's last major warehouse built in the old docklands is Kaispeicher A, designed by Werner Kallmorgen and completed in 1963. It rises 37m and stretches 126m along the River Elbe, but with a trapezoid-plan, in which its shortest side is just 22m wide. Plain brick facades are punctuated with small windows, and full-height vertical loading bays are on two sides. The city had plans to redevelop it, but a local developer approached Swiss architects Herzog and de Meuron to explore the idea of a concert hall retaining the distinctive warehouse. The city eventually agreed and the project became the Elbphilharmonie.

The architects designed a new volume clad with bespoke glass panels and with exactly the same plan as the warehouse

赫斯特大厦，福斯特合伙人事务所，美国纽约，2006年
Hearst Tower by Foster + Partners, New York, USA, 2006

部被完全掏空，只留下建筑的墙面。必须将650根桩子压低，才能为上方重达20万吨的扩建建筑提供支承力。旧体量目前用作停车场，但建筑师在其中设计了一条带有照明的隧道，并在隧道中安装了一部82m长、略微弯曲的电动扶梯，这样在一端看不到远处的另一端。这部扶梯可到达仓库建筑中唯一的一扇大窗户跟前，接着，在那里可以搭乘一部较短的电动扶梯到达"广场"。Kaispeicher A仓库的体量中包括了第三个音乐厅，即用于实验音乐的礼堂和一个教育中心。

从"广场"可以通过一系列楼梯上升至扩建建筑，到达音乐厅，并越过几何结构复杂的几个楼层看到内部的面貌。各个门厅均受到建筑外围护结构的限制，而安装在外围护结构上的2200块烧结玻璃面板，其中几乎有600块向外弯曲，形成了16000m²的立面。

可容纳550人的小型长方形音乐厅的内部覆盖了曲面木板隔声墙。它与易北爱乐厅最引人注目的空间——可容纳2100人的大音乐厅截然不同。这是一个位于新体量的中心的有机上空空间，高25m，宽50m，内墙上安装了10000块由石膏面板组成的隔声板。多层次的座位从各个角度向下延伸，围绕在舞台四边，所以观众几乎是围绕着演出而落座的。而一个两层高的倒置蘑菇形声音反射器悬浮其上。

空间的和谐性、表面的多样化、对细节的关注和易北爱乐厅庞大的规模是如此激动人心。它堪称一部建筑力作。但从该建筑异想天开的波浪形屋顶可以看出，建筑师并不想让它看起来太过严肃。

beneath it. It contains two concert halls, a hotel and luxury flats and is topped with a wavy roof that reaches 102m up from the street at one end. Beneath it at the Kaispeicher A's roof level is a public space called the Plaza, with an open perimeter promenade around the Elphilharmonie's glass-walled entrance lobby inside.

It is one of the most ambitious twenty-first century buildings in Germany, but it has seen construction costs rise over seven-fold from original estimates since the construction began in 2007. The warehouse was completely gutted leaving only its walls. 650 piles had to be driven down to support the 200,000 tonne extension above. The old volume is now car parking, but within it there rises an illuminated tunnel through which an 82m-long escalator curves gently so that the far end is invisible. It lands at a the warehouse's only large window, then a shorter escalator reaches the Plaza. The Kaispeicher A volume includes a third auditorium for experimental music and an education center.

A web of stairs rises into it from the Plaza, leading towards the concert halls and revealing interior sightlines over several storeys within a complex geometry. Foyers are bounded by the perimeter, where 2,200 fritted glass panels, almost 600 of which curve outwards, constitute the 16,000m² facade.

The 550-capacity rectangular Small Concert Hall is lined with acoustic walls of curved wooden ridges. It is very different from the Elbphilharmonie's most spectacular space, the 2,100-capacity Grand Concert Hall. This is an organic void 25m high and up to 50m wide lined with 10,000 gypsum panels for acoustic baffling in the heart of the new volume. Multiple tiers of seating descend towards the stage on all sides, so the audience almost surrounds the performance. A two-storey inverted mushroom-shaped sound reflector is suspended above it.

The symphony of spaces, the variety of surfaces, the attention to detail and the sheer scale of Elbphilharmonie are breathtaking. It is an architectural tour-de-force. But with its whimsical wavy roof, it does not take itself too seriously. Herbert Wright

当时光跨越屋顶 When Time Jumps through the Roof

德国汉堡易北爱乐厅

Herzog & de Meuron

对于易北爱乐厅的场地Kaispeicher A仓库，汉堡的大多数人并不陌生，但却从未真正地关注它。如今，它将成为一个新的社会和文化中心，融入汉堡人民的日常生活，并将吸引来自世界各地的参观者。这座建筑综合体包括一座爱乐音乐厅、一座室内乐厅、餐馆、酒吧、一个可以欣赏汉堡及海港景色的全景露台、公寓、一家酒店和停车设施。这些不同的用途融合在一栋建筑中，使得建筑好似一座功能齐全的城市。而就像一座城市一样，Kaispeicher A仓库和易北爱乐厅这两种截然不同的建筑的叠加恰恰确保了令人兴奋的、多变的空间序列：一方面，是体现Kaispeicher A仓库与港口的关系的原始和复古感；另一方面，则是易北爱乐厅奢华优雅的世界。在两者之间有一处广阔的公共和私人空间，其特性和规模各不相同：Kaispeicher A仓库的大型露台扩展开来如同一个新型的公共广场，与建造在其上部的爱乐厅的内部世界相互呼应。它是一个将前台听众和演奏者融为一体的空间，而实际上这两个群体正代表了建筑的风格。这座爱乐厅的建筑类型经历了相当激进的建筑重组，它极为重视艺术家与听众之间的接近程度，就好像一座足球场。

这座新建的爱乐厅不仅是一个音乐圣地，还是一个成熟的居住和文化综合设施。可以容纳2100人的音乐厅和可以容纳550名听众的室内乐厅融入豪华公寓和一家配备了餐馆、运动健身中心和会议设施等服务设施的五星级酒店内部。作为第二次世界大战结束后长期默默无闻的一处古迹，Kaispeicher A仓库偶尔举办一些小型活动，而如今，它被改造成了一个充满活力的国际音乐爱好者中心，吸引了来自旅游者和商界的关注。

由维尔纳·卡尔摩根担任设计师的Kaispeicher A仓库，建于1963年至1966年之间，并在20世纪末停止使用之前一直被当作仓库。该仓库的实体结构最初建造用于储存成千上万袋沉重的可可豆，而目前用于为新爱乐厅提供支撑力。19世纪港口仓库的设计都会呼应城市中历史悠久的立面风格：仓库的窗户、地基、山墙和许多装饰性构件都与当时的建筑风格保持一致。从易北河畔望去，尽管这些建筑物是不适宜居住的仓库，既不要求也不需要引入光线、空气和阳光，但是它们的设计也有意识地融入了城市的天际线。然而，Kaispeicher A仓库是例外：虽然它与汉堡港口的其他许多仓库一样也是一栋构造结实的巨大砖混建筑，但其陈旧的立面设计风格抽象并超然不俗。该建筑立面洞口组成的规则网格（尺寸为50cm×75cm）几乎不能被称为窗户，它们更像是结构而非洞口设计。

新建筑沿着Kaispeicher A仓库的外形向上伸出；它与砖砌旧建筑的平面设计保持一致，矗立于旧建筑之上。然而，新结构的顶部和底部采用了与下方仓库平静而简易的外形截然不同的策略：屋顶起伏的

下摆从较低的东端向上升高，在Kaispitze (半岛结构的尖端)达到108m的高度。

　　易北爱乐厅是一栋从很远处就看得到的地标性建筑，为汉堡这座以平面布局为特征的城市增加了一栋全新的突出垂直线条的建筑。在这个崭新的城市空间内，宽阔的水面和远洋轮的工业规模让此处产生了更加强烈的空间感。一部分由弧形面板组成的玻璃立面，一些采用切口设计，将位于旧建筑顶部的新建筑变成一块巨大的、彩虹色的水晶，随着它对天空、水面和城市倒影的捕捉，它的外表也在不断地发生变化。上层建筑的下部也带有富有表现力的动态效果。人们可以从沿着建筑的边缘设计的拱形洞口中看到天空，拱形洞口打造出了易北河和汉堡市中心壮观且戏剧化的风景。在建筑内部更深处，深深的垂直洞口在广场和不同楼层的门厅之间制造了不断变化的视觉联系。

　　Kaispeicher A仓库综合体的主入口位于东部。搭乘一部超长的自动扶梯可以通往广场，广场带有轻微曲线，所以从广场的一端看不到另一端的全貌。它直通整个Kaispeicher A仓库，经过一扇带有阳台的大型全景窗户，在阳台上可以欣赏到港口的景色，然后继续往上走可到达广场。而广场位于Kaispeicher A仓库的顶部，也就是新建筑的下方，就像是新旧建筑之间一个巨大的折叶。它是一个可以欣赏独特全景的新公共空间。

Elbphilharmonie, Hamburg

The Elbphilharmonie on the Kaispeicher A marks a location that most people in Hamburg know about but have never really noticed. It is now set to become a new center of social, cultural and daily life for the people of Hamburg and for visitors from all over the world. The building complex accommodates a philharmonic hall, a chamber music hall, restaurants, bars, a panorama terrace with views of Hamburg and the harbour, apartments, a hotel and parking facilities. These varied uses are combined in one building as they are in a city. And like a city, the two contradictory and superimposed architectures of the Kaispeicher A and the Philharmonic ensure exciting, varied spatial sequences: on the one hand, the original and archaic feel of the Kaispeicher A marked by its relationship to the harbour; on the other, the sumptuous,

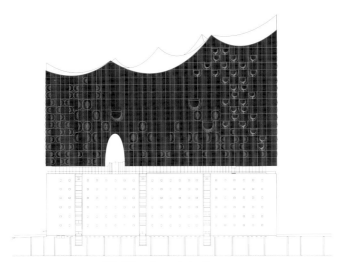

西北立面 north-west elevation

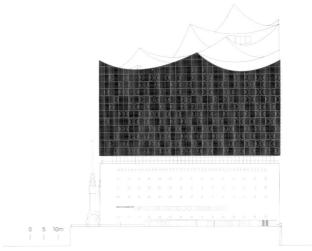

东北立面 north-east elevation

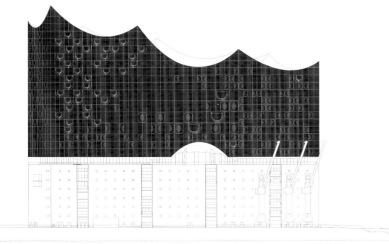

东南立面 south-east elevation

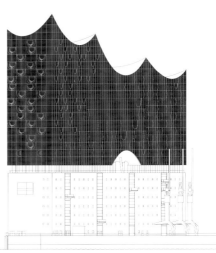

西南立面 south-west elevation

elegant world of the Elbphilharmonic. In between, there is an expansive topography of public and private spaces, all differing in character and scale: the large terrace of the Kaispeicher A, extending like a new public plaza, responds to the inwardly oriented world of the Elbphilharmonic built above it. A space has emerged that foregrounds music listeners and music makers to such an extent that, together, they actually represent the architecture. The philharmonic building typology has undergone architectural reformulation that is exceptionally radical in its unprecedented emphasis on the proximity between artist and audience – almost like a football stadium.

The new philharmonic is not just a site for music; it is a full-fledged residential and cultural complex. The concert hall, seating 2100, and the chamber music hall for 550 listeners are embedded in between luxury flats and a five-star hotel with built-in services such as restaurants, a health and fitness

1 KAISPEICHER The old Kaispeicher A with its red brick façade is the foundation that the Elbphilharmonie is built upon. At the start of the construction process, the old warehouse for cocoa, tea and tobacco was gutted completely.

2 THE FAÇADE The glass façade consists of 1,100 individual window panels, intricately curved and with a pattern of individually printed grey chrome points. The entire surface of the façade is equal in area to two football fields.

3 THE TUBE At the end of the gently arched 82 metre escalator a panorama window awaits the visitor, offering a view of the harbour. Another 20 metre escalator goes up to the Plaza.

4 THE PLAZA The central platform, at a height of 37 metres, is an open space, accessible to the public. The outside promenade, circling the whole building, offers spectacular views of the harbour and the city skyline. The Plaza covers an area of 4,000 square metres and is about as large as Hamburg's Town Hall Square.

5 THE GRAND HALL The heart of the Elbphilharmonie is the Grand Hall. With a seating capacity of 2,100, the concert hall is structured like a vineyard, with a stage at the centre, surrounded by terrace-like balconies for the audience.

6 THE SOUND REFLECTOR A large sound reflector is suspended from the centre of the vaulted ceiling and guarantees excellent acoustics. The sound is evenly distributed around the concert hall by the reflector.

7 THE ORGAN The four-manual organ, with 65 stops and additional stops located within the sound reflector on the ceiling, completes the Grand Hall.

8 THE RECITAL HALL On the east side of the building, a Recital Hall with a flexible stage and variable seating offers places for up to 550 people.

9 THE KAISTUDIO The Kaistudio is located within the former warehouse structure. With 150 seats, it is designed to be a place for experimental music, lectures and workshops. Above all, however, the Kaistudio is the heart of the music education area which extends over two floors of the Elbphilharmonie and provides a new home for the Klingende Museum, the interactive instrument museum.

10 THE FOYER BAR The foyer bar on the 15th floor is the gastronomic centre within the concert hall. There are additional bar areas for guests to meet and relax during the intervals.

11 HOTEL Located on the east side of the building is a 14-storey hotel with 250 guest rooms and a spa and conference centre.

12 RESIDENTIAL APARTMENTS 45 spacious apartments with glass fronts and balconies offer spectacular views of the river Elbe, the harbour and the city.

13 CAR PARK Through the entrance on the east side of the building, a spiral ramp leads up to the seven-storey car park in the lower part of the building. The garage can accommodate more than 500 vehicles.

项目名称：Elbphilharmonie, Hamburg / 地点：Platz der Deutschen Einheit 1-5, Hamburg, Germany / 建筑师：Herzog & de Meuron / 合伙人：Jacques Herzog, Pierre de Meuron, Ascan Mergenthaler_Partner in Charge, David Koch_Partner in Charge Project Management / 项目团队：Jan-Christoph Lindert, Nicholas Lyons, Stefan Goeddertz, Christian Riemenschneider, Henning Severmann, Stephan Wedrich, Carsten Happel, Birgit Föllmer, Kai Zang, Peter Scherz, Jan Per Grosch / 总设计师：Herzog & de Meuron GmbH, H+P Planungsgesellschaft mbH & Co. KG, Hochtief Solutions AG / 电气/结构工程：Hochtief Solutions AG / 标识设计：Herzog & de Meuron GmbH with Integral Ruedi Baur, Hochtief Solutions AG / 洒水器供应商：Itega GmbH Ingenieurbüro für technische Gebäudeausrüstung. Hochtief Solutions AG / 声效设计：Nagata Acoustics Inc. / 建筑物理：MF Dr. Flohrer Beratende Ingenieure GmbH, Hochtief Solutions AG

垂直交通：Jappsen Ingenieure GmbH / 防火、工地监管：Hahn Consult Ingenieurgesellschaft / 人流通过设计：Happold Ingenieurbüro GmbH, Arbeitsgemeinschaft Planung Elbphilharmonie / 立面维护策略：Univ.-Prof. Dr.-Ing. Manfred Helmus Ingenieurpartnerschaft / 噪声控制：Taubert und Ruhe GmbH / 风力工程顾问：Wacker Ingenieure / 立面工程：R+R Fuchs / 气候顾问：Transsolar / 设施管理：SPIE GmbH / 住宅室内设计师：Antonio Citterio and Partners / 照明设计合作方：Ulrike Brandi Licht / 照明设计：ARGE Planung Elbphilharmonie, ARGE Generalplaner Elbphilharmonie / 客户：Freie und Hansestadt Hamburg / 用地面积：10,540m² / 建筑面积：5,745m² / 总建筑面积：125,512m² / 楼层数量：29 / 设计时间：2003.4—2006.6 / 施工时间：2007.4—2016.10 / 摄影师：©Iwan Baan (courtesy of the architect)

二层 second floor

1. 主入口	20. 广场上空空间	1. main entrance	19. chamber music hall
2. 自动扶梯	21. 后台	2. escalator	20. void of plaza
3. 售票处	22. 指挥室	3. ticketing	21. backstage
4. 通道	23. 独奏者房间	4. passageway	22. conductor's room
5. 停车场	24. 艺术总监房间	5. parking	23. soloist's room
6. 酒店入口	25. 酒店	6. entrance of hotel	24. artistic director
7. 住宅入口	26. 主要音乐厅	7. entrance of residential	25. hotel
8. 卸货区	27. 室内乐厅上空空间	8. loading bays	26. main concert hall
9. 酒店驱动装置	28. 休息室	9. hotel drive	27. airspace of chamber music hall
10. 电梯厅	29. 校音室	10. lift lobby	28. lounge
11. 表演工作室1号	30. 住所	11. performance studio "Kaistudio 1"	29. tuning room
12. 表演工作室2号	31. 控制室	12. performance studio "Kaistudio 2"	30. residential
13. 门厅	32. 管风琴	13. foyer	31. control room
14. 酒店管理办公室	33. 酒店上空空间	14. hotel administration	32. organ
15. 广场	34. 屋顶露台	15. plaza	33. void of hotel
16. 酒店大堂	35. 住宅上空空间	16. hotel lobby	34. roof terrace
17. 商店		17. shop	35. void of residential
18. 咖啡厅		18. cafe	
19. 室内乐厅			

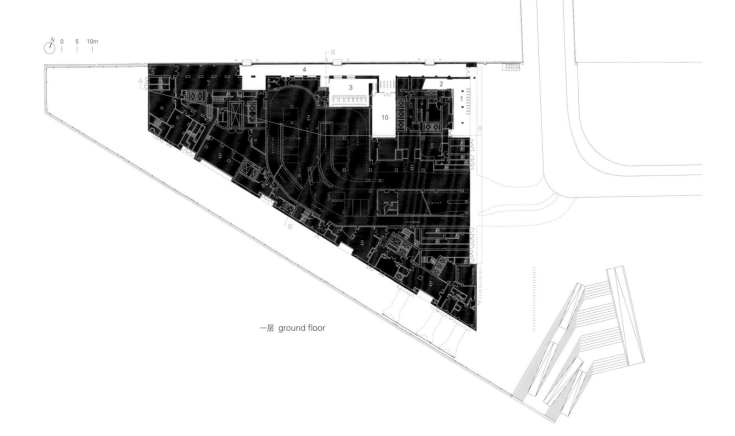

一层 ground floor

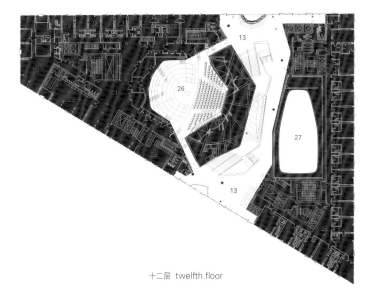

十二层 twelfth floor

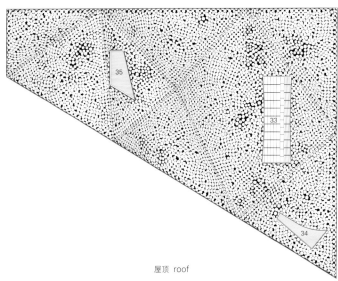

屋顶 roof

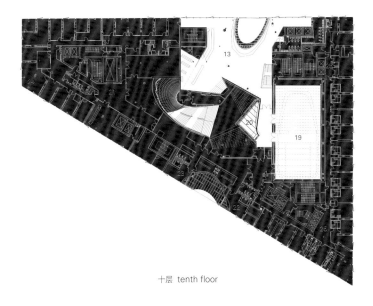

十层 tenth floor

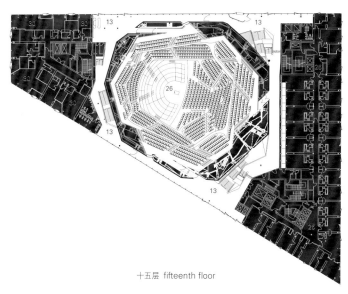

十五层 fifteenth floor

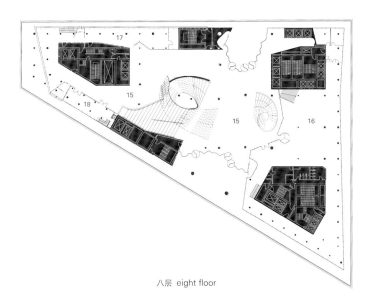

八层 eight floor

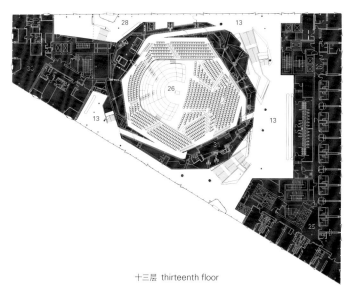

十三层 thirteenth floor

center, conference facilities. Long a mute monument of the post-war era that occasionally hosted fringe events, the Kaispeicher A has now been transformed into a vibrant, international center for music lovers, a magnet for both tourists and the business world.

The Kaispeicher A, designed by Werner Kallmorgen, was constructed between 1963 and 1966 and used as a warehouse until close to the end of the last century. Originally built to bear the weight of thousands of heavy bags of cocoa beans, it now lends its solid construction to supporting the new philharmonic. The harbour warehouses of the 19th century were designed to echo the vocabulary of the city's historical façades: their windows, foundations, gables and various decorative elements are all in keeping with the architectural style of the time. Seen from the River Elbe, they were meant to blend in with the city's skyline despite the fact that they were uninhabited storehouses that neither required nor invited the presence of light, air and sun. But they are not the Kaispeicher A: it is a heavy, massive brick building like many other warehouses in the Hamburg harbour, but its archaic façades are abstract and aloof. The building's regular grid of holes measuring 50 x 75 cm cannot be called windows; they are more structure than opening.

The new building has been extruded from the shape of the Kaispeicher A; it is identical in the ground plan with the brick block of the older building, above which it rises. However, at the top and bottom, the new structure takes a different tack from the quiet, plain shape of the warehouse below: the

undulating sweep of the roof rises from the lower eastern end to its full height of 108 meters at the Kaispitze (the tip of the peninsula). The Elbphilharmonie is a landmark visible from afar, lending an entirely new vertical accent to the horizontal layout that characterises the city of Hamburg. There is a greater sense of space here in this new urban location, generated by the expanse of the water and the industrial scale of the seagoing vessels. The glass façade, consisting in part of curved panels, some of them carved open, transforms the new building, perched on top of the old one, into a gigantic, iridescent crystal, whose appearance keeps changing as it catches the reflections of the sky, the water and the city. The bottom of the superstructure also has an expressive dynamic. Along its edges, the sky can be seen from the Plaza through vault-shaped openings, creating spectacular, theatrical views of both the River Elbe and downtown Hamburg. Further inside, deep vertical openings provide ever-changing visual relations between the Plaza and the foyers on different levels. The main entrance to the Kaispeicher A complex lies to the east. An exceptionally long escalator leads up to the Plaza; it describes a slight curve so that it cannot be seen in full from one end to the other. It cuts straight through the entire Kaispeicher A, passing a large panorama window with a balcony that affords a view of the harbour before continuing on up to the Plaza. The latter, sitting on top of the Kaispeicher A and under the new building, is like a gigantic hinge between old and new. It is a new public space that offers a unique panorama. Herzog & de Meuron

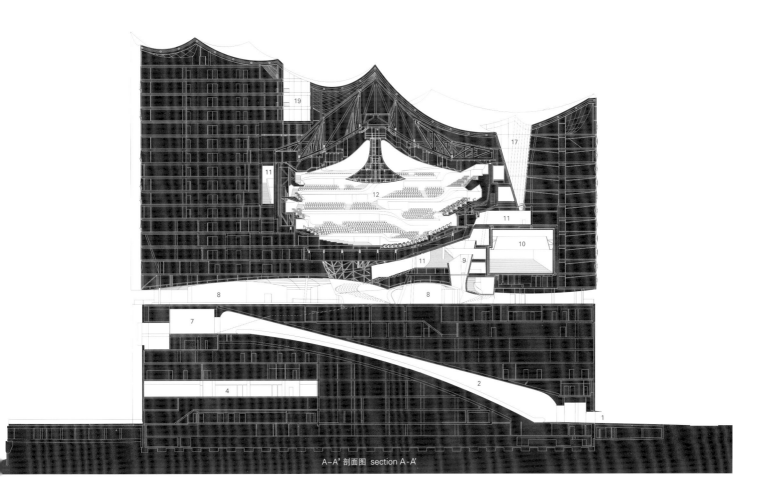

A-A' 剖面图 section A-A'

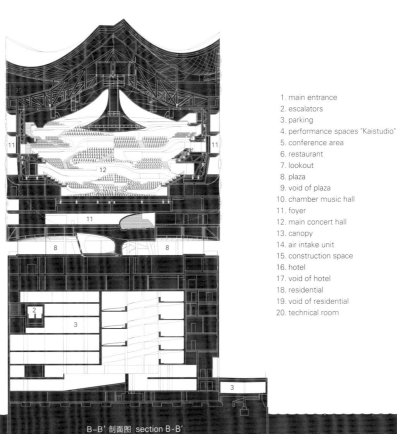

B-B' 剖面图 section B-B'

1. 主入口
2. 自动扶梯
3. 停车场
4. Kaistudio表演空间
5. 会议室
6. 餐厅
7. 瞭望台
8. 广场
9. 广场上空空间
10. 室内乐厅
11. 门厅
12. 主要音乐厅
13. 雨篷
14. 进风口
15. 构造空间
16. 酒店
17. 酒店上空空间
18. 住宅
19. 住宅上空空间
20. 机械设备间

1. main entrance
2. escalators
3. parking
4. performance spaces "Kaistudio"
5. conference area
6. restaurant
7. lookout
8. plaza
9. void of plaza
10. chamber music hall
11. foyer
12. main concert hall
13. canopy
14. air intake unit
15. construction space
16. hotel
17. void of hotel
18. residential
19. void of residential
20. technical room

当时光跨越屋顶 When Time Jumps through the Roof

安特卫普港口之家
Zaha Hadid Architects

位于安特卫普的这座新建的港口之家通过将一处废弃的消防站转换用途、改造和扩建,将其变成港口的新总部,从而将先前工作于城市各处办公大楼中的500名港口员工聚集于一处。

安特卫普的码头长达12km,是欧洲第二大海运港口,它每年为15000艘海上贸易船只和60000艘内河驳船提供服务。安特卫普掌握26%的欧洲集装箱运输,通过停靠在港口的航运船只运输2亿多吨货物,并为60000多人提供就业机会,其中包括8000多名港口工人。

2007年,当先前建于20世纪90年代的安特卫普港口办公室无法满足需求时,港口管理层决定只有迁移办公地点才能实现技术和行政事务的结合,为大约500名员工提供新的办公场所。

作为城市和巨大港口之间的门槛,位于安特卫普63号码头Kattendijk船坞的墨西哥岛被选为新总部的场地。同时,这块滨水场地也提供了显著的可持续性施工效益,使得建筑材料和建筑构件可以通过水路运输,极大地满足了港口的生态目标。

随着新消防站的建造,其配套设施能够满足扩建港口所需的服务,位于墨西哥岛场地的旧消防站(一栋仿汉萨同盟住所而建的列管建筑)就变得多余了,所以必须改变其用途才能确保这座建筑得到保护,建筑师必须将这一废弃的消防站整合进新项目当中。

通过与历史古迹修复和改造领域的龙头顾问公司Origin的合作，扎哈·哈迪德建筑师事务所（ZHA）对场地历史和古迹的研究成为设计的基础，首先将重点放在平行于Kattendijk船坞的南北场地轴线上，将城市中心与港口连接。其次，鉴于场地环水的地理位置，建筑的四个立面具备同等的重要性，没有主次立面之分。ZHA采用在旧建筑上方扩建的方式，而不是在旁边扩建，也就不会遮挡任何一个现有立面了。ZHA和Origin对旧消防站的历史分析还突出了其原先打算设计的塔楼的作用，也就是把它当成具有汉萨同盟设计风格的消防站的一个宏伟壮观的组成部分。可惜其大胆的垂直设计意图一直未得到实现，原本想要成为下方建筑体量的顶部。

这三个关键原则定义了设计中新和旧的组成：新体量"漂浮于"旧建筑之上，尊重每一个旧的立面，并实现了初始设计中未能实现的塔楼的垂直特性。新扩建建筑好似一艘船的船头，指向斯凯尔特河，将建筑与安特卫普赖以为生的这条河流联系在一起。

新扩建建筑三面环水，立面采用玻璃表面，如波浪起伏一般闪烁，并反射不断变化的城市天空的颜色。三角形玻璃板的应用使得在建筑任意一端都能形成明显光滑的曲线形体，这些玻璃面板还有助于实现由建筑南端的平直立面渐变为北端波纹状的表面。

尽管大多数三角形面板都是透明的，但也有一些采用不透明的设计。此种标准化的混合设计确保建筑内部拥有充足的阳光，并且同时控制太阳负荷以保证最优化的工作环境。同时，透明和不透明立面板材的交替变化打破了新扩建建筑的体量整体感，并可饱览斯凯尔特河、城市和港口的景色，同时也为建筑提供了外围护结构。

在建筑南端，用平玻璃板形成了立面的波纹状特性，再逐渐过渡至北侧更加立体的设计。这种透明体量设计赋予新建筑闪闪发光的外表，重新诠释了安特卫普"钻石之城"的称号。新扩建建筑呈现出一种精心切割的外形，它能够随着日光强度的变化而变化。好似周围港口水面上的波纹一样，新立面反映了不断变化的光照情况。

建筑师用一个玻璃屋顶将旧消防站的中央庭院封闭，将其变成新港口之家的主接待区。经由该中庭，访客可以进入位于经过精心修复和保护的废弃消防车大厅中的具有历史性意义的公共阅览室和图书馆。搭乘观光电梯可以直接进入带有室外人行桥的新扩建建筑，该桥梁连接现有建筑和新扩建建筑，并可以一览城市和港口的景色。

客户要求在设计中融入"活动式办公室"的概念，并带有位于现有建筑上层中心和新扩建建筑底部楼层的餐厅、会议室和礼堂等相关区域。而与中心相距较远的其他楼层由开放式办公室组成。

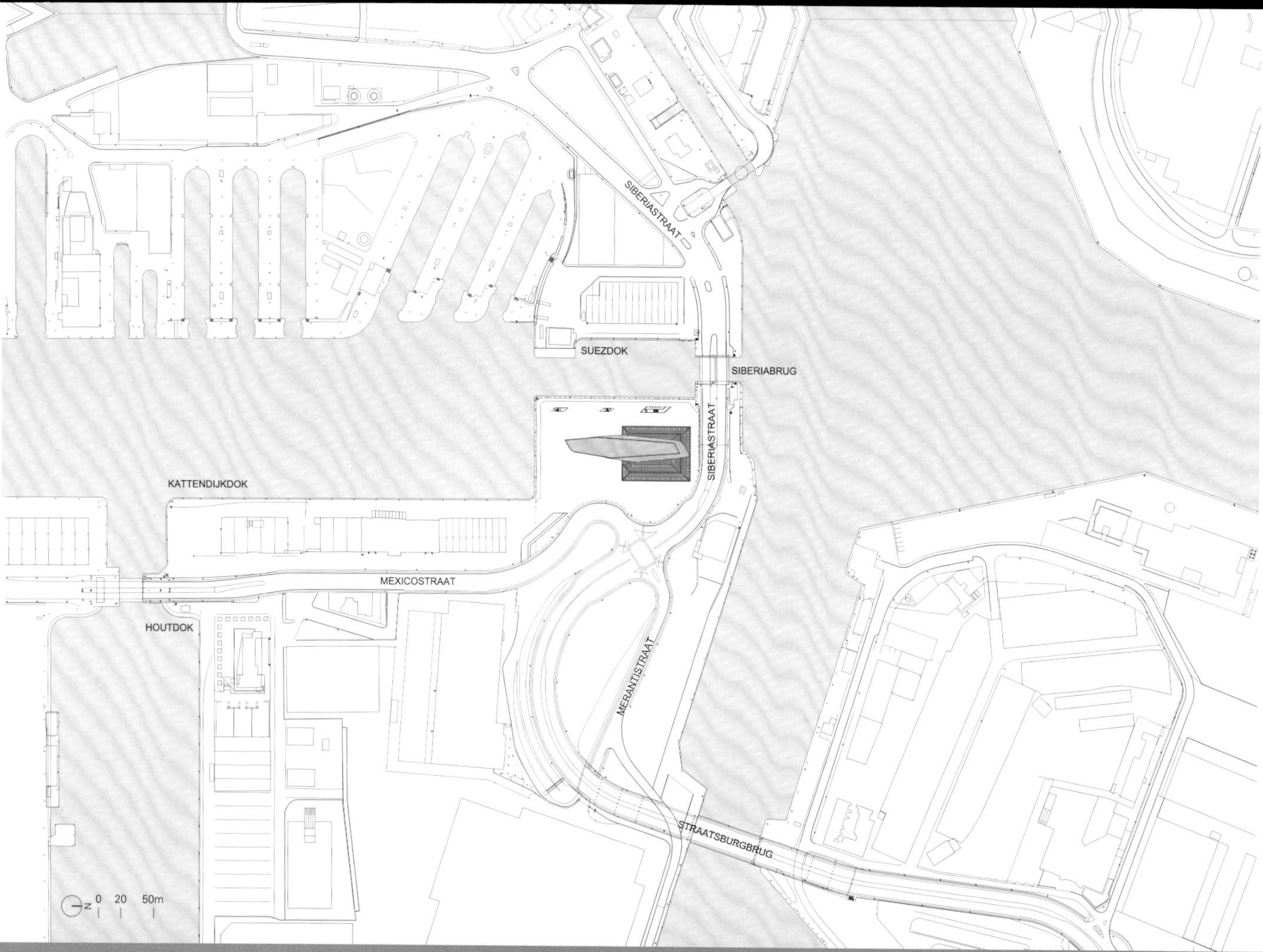

KATTENDIJKDOK
SUEZDOK
SIBERIABRUG
SIBERIASTRAAT
MEXICOSTRAAT
HOUTDOK
MERANTISTRAAT
STRAATSBURGBRUG

0 20 50m

Port House in Antwerp

The new Port House in Antwerp repurposes, renovates and extends a derelict fire station into a new headquarters for the port – bringing together the port's 500 staff that previously worked in separate buildings around the city.

With 12 km of docks, Antwerp is Europe's second largest shipping port, serving 15,000 sea trade ships and 60,000 inland barges each year. Antwerp handles 26% of Europe's container shipping, transporting more than 200 million tonnes of goods via the ocean-going vessels that call at the port and providing direct employment for over 60,000 people, including more than 8,000 port workers.

In 2007, when the former 1990s offices of the Port of Antwerp had become too small, the port determined that relocation would enable its technical and administrative services to be housed together, providing new accommodation for about 500 staff.

As the threshold between the city and its vast port, Mexico Island in Antwerp's Kattendijk dock on Quay 63 was selected as the site for the new head office. The waterside site also offered significant sustainable construction benefits, allowing materials and building components to be transported by water, an important requirement to meet the port's ecological targets.

Following the construction of a new fire station with facilities needed to service the expanding port, the old fire station on the Mexico Island site – a listed replica of a Hanseatic residence – became redundant and relied on a change of use to ensure its preservation. This disused fire station had to be integrated into the new project.

Working with Origin, leading heritage consultants in the restoration and renovation of historic monuments, ZHA's studies of the site's history and heritage are the foundations of the design which firstly emphasises the north-south site axis parallel with the Kattendijk dock linking the city center to

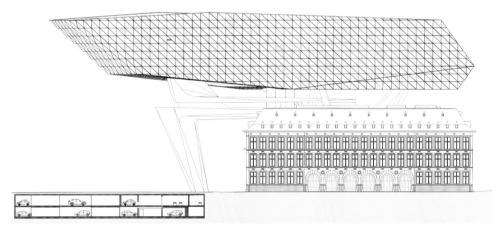

东立面 east elevation

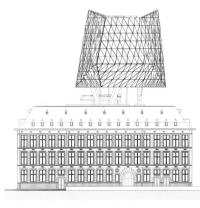

北立面 north elevation

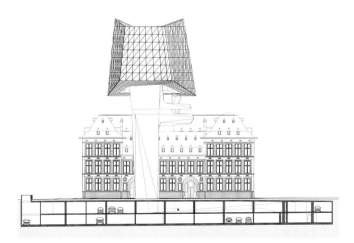

南立面 south elevation

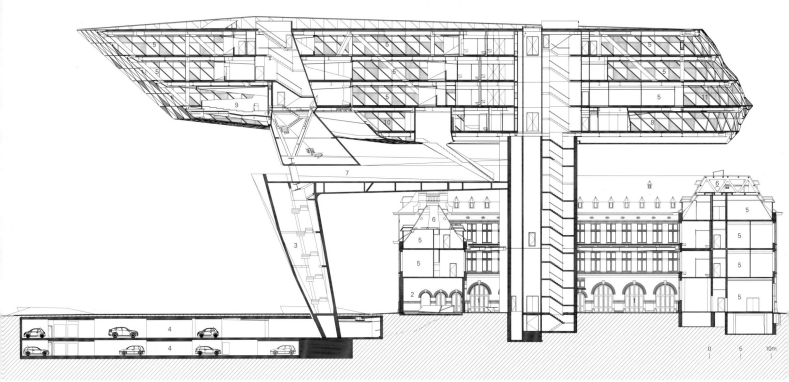

A-A' 剖面图 section A-A'

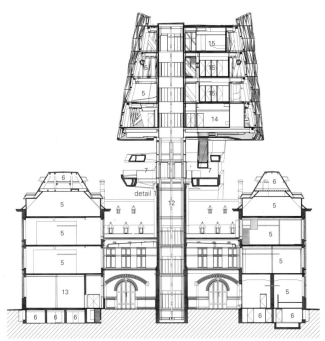

B-B' 剖面图 section B-B'

1. 庭院	1. courtyard
2. 主入口大厅	2. main entrance hall
3. 楼梯	3. stairs
4. 停车场	4. parking
5. 办公室/会议室	5. offices / meeting room
6. 技术设备区	6. technical area
7. 观景台	7. viewing deck
8. 餐厅	8. restaurant
9. 礼堂	9. auditorium
10. 门厅	10. foyer
11. 董事会会议室	11. boardroom
12. 电梯核心筒	12. lift core
13. 图书馆	13. library
14. 厨房	14. kitchen
15. 卫生间	15. toilets

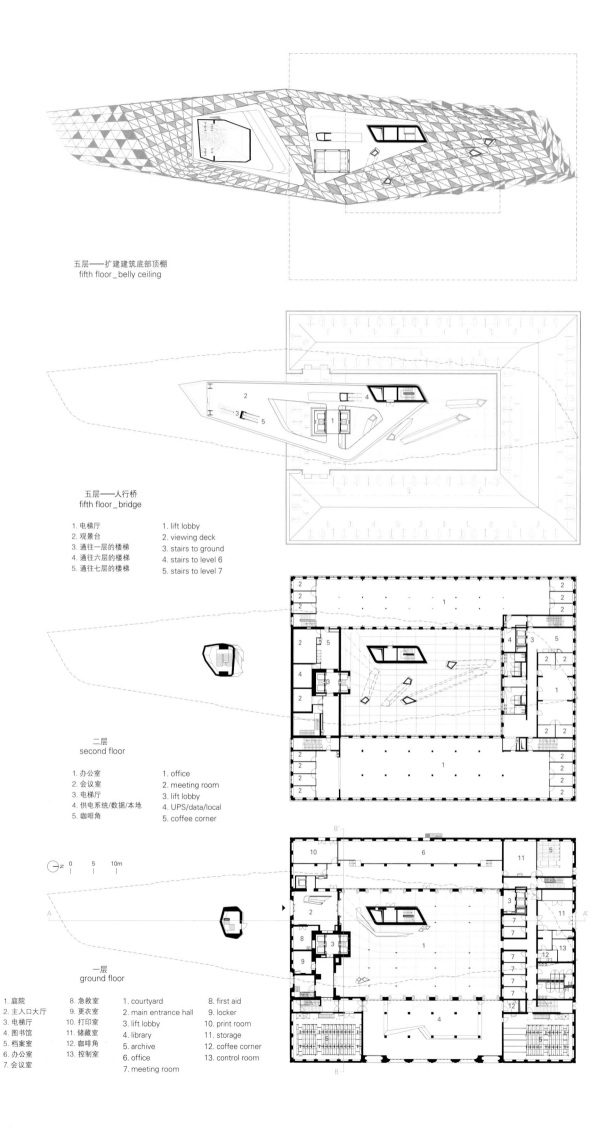

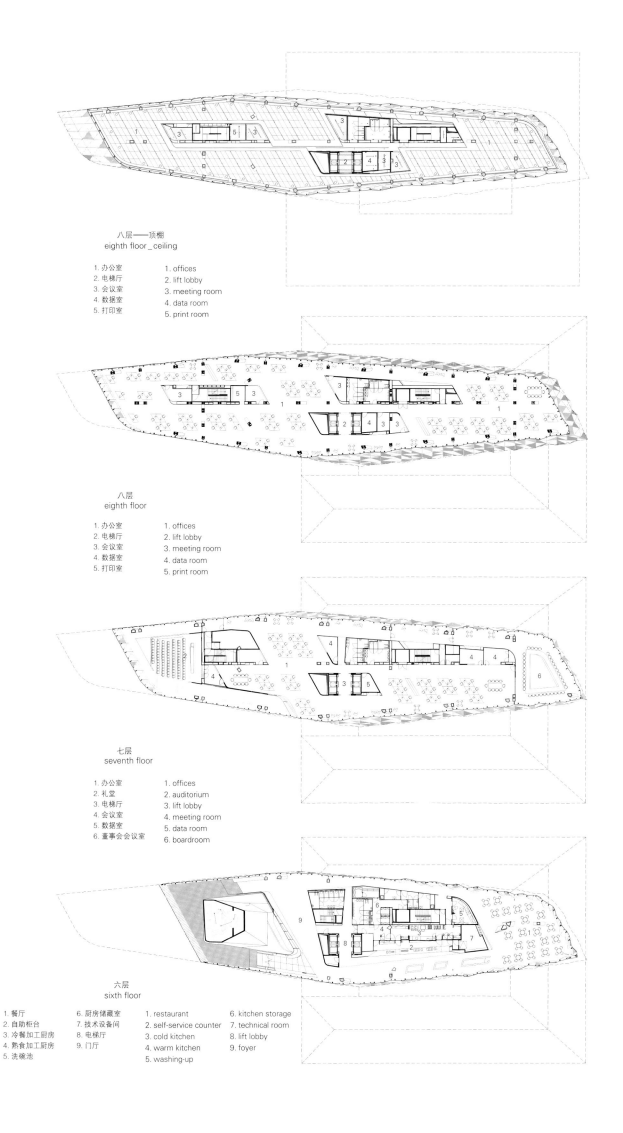

八层——顶棚
eighth floor_ceiling

1. 办公室
2. 电梯厅
3. 会议室
4. 数据室
5. 打印室

1. offices
2. lift lobby
3. meeting room
4. data room
5. print room

八层
eighth floor

1. 办公室
2. 电梯厅
3. 会议室
4. 数据室
5. 打印室

1. offices
2. lift lobby
3. meeting room
4. data room
5. print room

七层
seventh floor

1. 办公室
2. 礼堂
3. 电梯厅
4. 会议室
5. 数据室
6. 董事会会议室

1. offices
2. auditorium
3. lift lobby
4. meeting room
5. data room
6. boardroom

六层
sixth floor

1. 餐厅
2. 自助柜台
3. 冷餐加工厨房
4. 熟食加工厨房
5. 洗碗池
6. 厨房储藏室
7. 技术设备间
8. 电梯厅
9. 门厅

1. restaurant
2. self-service counter
3. cold kitchen
4. warm kitchen
5. washing-up
6. kitchen storage
7. technical room
8. lift lobby
9. foyer

the port. Secondly, due to its location surrounded by water, the building's four elevations are considered of equal importance with no principal facade. ZHA's design is an elevated extension, rather than a neighbouring volume which would have concealed at least one of the existing facades. ZHA and Origin's historic analysis of the old fire station also highlighted the role of its originally intended tower – a grand, imposing component of the fire station's Hanseatic design. Its bold vertical statement, intended to crown the imposing volume of the building below, was never realised.

These three key principles define the design's composition of new and old: a new volume that "floats" above the old building, respecting each of the old facades and completing the verticality of the original design's unrealised tower. Like the bow of a ship, the new extension points towards the Scheldt, connecting the building with the river on which Antwerp was founded.

Surrounded by water, the new extension's facade is a glazed surface that ripples like waves and reflects the changing tones and colours of the city's sky. Triangular facets allow the apparently smooth curves at either end of the building to be formed with flat sheets of glass. They also facilitate the gradual transition from a flat facade at the south end of the building to a rippling surface at the north.

While most of the triangular facets are transparent, some are opaque. This calibrated mix ensures sufficient sunlight within the building, while also controlling solar load to guarantee optimal working conditions. At the same time, the alternation of transparent and opaque facade panels breaks down the volume of the new extension, giving panoramic views of the

Scheldt, the city and the port as well as providing enclosure. The facade's rippling quality is generated with flat facets to the south that gradually become more three-dimensional towards to the north. This perception of a transparent volume, cut to give the new building its sparkling appearance, reinterprets Antwerp's moniker as "the city of diamonds". The new extension appears as a carefully cut form which changes its appearance with the shifting intensity of daylight. Like the ripples on the surface of the water in the surrounding port, the new facade reflects changing light conditions.

The old fire station's central courtyard has been enclosed with a glass roof and is transformed into the main reception area for the new Port House. From this central atrium, visitors access the historic public reading room and library within the disused fire truck hall which has been carefully restored and preserved. Panoramic lifts provide direct access to the new extension with an external bridge between the existing building and new extension giving panoramic views of the city and the port.

The client requirements for an "activity based office" are integrated within the design, with related areas such as the restaurant, meeting rooms and auditorium located at the center of the upper levels of the existing building and the bottom floors of the new extension. The remaining floors more remote from the center, comprise open plan offices.

86

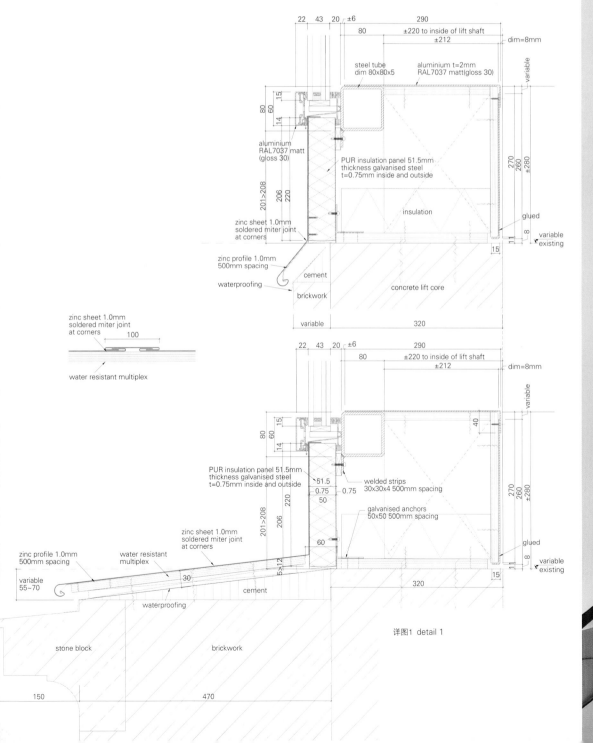

详图1 detail 1

项目名称：Port House in Antwerp / 地点：Antwerp, Belgium / 建筑师：Zaha Hadid Architects (ZHA) / 设计师：Zaha Hadid and Patrik Schumacher / 项目总监：Joris Pauwels / 项目建筑师：Jinmi Lee / 项目团队：Florian Goscheff, Monica Noguero, Kristof Crolla, Naomi Fritz, Sandra Riess, Muriel Boselli, Susanne Lettau / 竞赛团队：Kristof Crolla, Sebastien Delagrange, Paulo Flores, Jimena Araiza, Sofia Daniilidou, Andres Schenker, Evan Erlebacher, Lulu Aldihani / 执行建筑师：Bureau Bouwtechniek / 结构工程师：Studieburo Mouton Bvba / 设备工程师：Ingenium Nv / 声效工程师：Daidalos Peutz / 修复顾问：Origin / 防火工程：Fpc / 用地面积：16,400m² / 总建筑面积：12,800m² (6,600m² in the refurbished fire station, 6,200m² in the new extension) / 建筑规模：111m length x 24m width x 21m height (new extension); 63m length x 78.5m width x 21.5m height (existing fire station) Total height (existing building + new extension): 46m (5 additional floors) / 竣工时间：2016 / 摄影师：©Helene Binet (courtesy of the architect) - p.76~77, p.78, p.81; ©Hufton + Crow (courtesy of the architect) - p.74, p.85, p.87, p.88~89; ©Tim Fisher (courtesy of the architect) - p.72~73, p.79, p.84, p.86

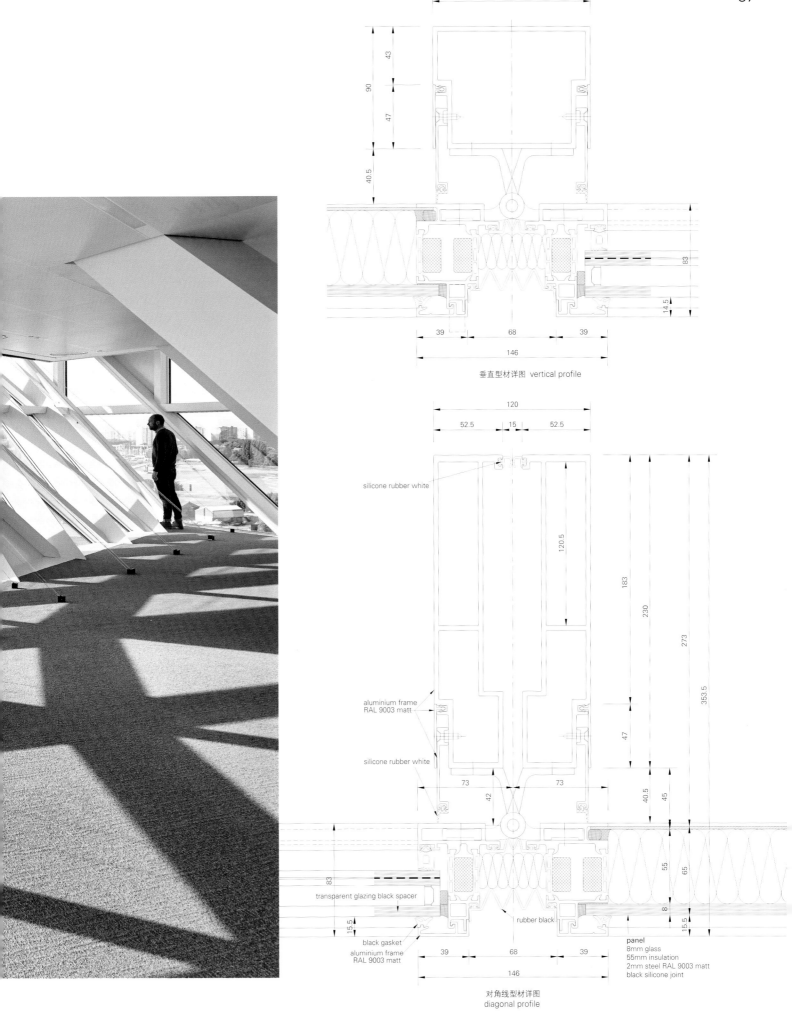

麦克唐纳贸易中心

Aerial view, 2010

仓库屋顶为巴黎警察局的拘留所 roof of the warehouse occupied by the pound of the Paris Police Office

麦克唐纳贸易中心的北立面,1970年
north facade of the Entrepôt Macdonald, 1970

N3大楼的北立面视图,2016年
north facade in 2016, view of the building N3

O1大楼的西南立面,从项目的一端看去 south-west facade, view from the end of the building O1

麦克唐纳贸易中心是一个经15名建筑师设计、由23个项目组成、独立的城市街区。项目建造在一座占地142200m²的超级建筑之上，它是一座完成于1970年的三层仓库，是巴黎最长的建筑。

2001年，巴黎发布了该市的"旧城改造大项目（GPRU）"，确定了位于城市边缘的工业和交通场地需要进行改造和住宅开发。在巴黎东北部，麦克唐纳贸易中心占GPRU潜在建筑总量的15%。未来将建造一座新的RER（区域特快列车）车站——罗莎·帕克斯站。2006年，麦克唐纳贸易中心被出售给了SAS ParisNordEst组织，该组织成立的目的是交付带有配套基础设施的GPRU住宅项目。

2007年，OMA的雷姆·库哈斯和弗洛里斯·阿克梅德赢得了麦克唐纳贸易中心总体规划设计竞赛。阿克梅德构思了该设计方案的指导方针：尊重现有的纪念性质和上层建筑；保留仓库的直线性，但设计了一条从其中心穿过的有轨电车路线；在屋顶建造新建筑；引入自然光线；实现多种用途；并且保留617m长的北侧立面一楼最初的混凝土立面。研究证实，混凝土框架及其8m×8.5m的柱网十分坚固，足以为新扩建建筑（包括屋顶的建筑）提供支承力。对旧建筑的重新使用既可以保留内部固有的能量，同时还能防止污染排放。该总体规划实现了165000m²的多功能空间（或包括停车场在内的220000m²的空间）。2008年，15名建筑师接受委托设计场地中的不同部分，设计必须符合立面的水晶（透明的）和矿石（不透明的）水平分层要求。

麦克唐纳贸易中心包括面积超过74300m²的1130套住宅（其中641套为社会福利住宅）、一个占地16000m²的企业孵化园、占地28000m²的办公和首层商铺，还有占地16300m²的大学、小学、运动中心、幼儿园和社区中心。在地面以上一层，仓库被分成两个平行的六面体，中间有花园和一连串的庭院。经由一条18.5m宽的北侧散步路可以向下通往路边，而南侧正对着罗莎·帕克斯车站。

基础设施工程开始于2010年，而结构工程开始于2012年。电车于2012年12月投入使用。办公室项目竣工于2014年，大学也于同年对外开放。最后建造的CARGO企业孵化园于2015年11月完工。商铺于2016年年初对外开放，而公共空间的装修也于同一时期完工。如今，麦克唐纳贸易中心为3500户居民、4000名工人和1000名学生提供了住所和办公、学习场所。

Entrepôt Macdonald

The Entrepôt Macdonald is a self-contained city neighbourhood which incorporates 23 projects by 15 architects. It is built on the base of a single 142,200 m² megastructure – a three-storey warehouse completed in 1970, the longest building in Paris.

In 2001, Paris published its Grand Urban Renewal Project (GPRU), which identified industrial and transport sites at the city edge, for transformation and residential development. In north-east Paris, the Entrepôt Macdonald had 15% of the GPRU's building potential. A new RER (regional express train) station, Rosa Parks, would be built. In 2006, the Entrepôt Macdonald was sold to SAS ParisNordEst, established to deliver the GPRU housing with supporting infrastructure.

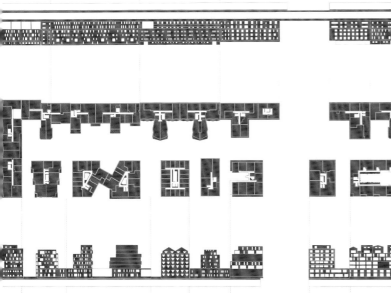

项目名称：Entrepôt Macdonald
地点：141-221 Boulevard Macdonald, 19th arrondisement, Paris, France
总体规划和协调建筑师：Floris Alkemade in association with Xaveer De Geyter
项目经理：Milena Wysoczynska (XDGA)
建筑师：Kengo Kuma, Studio Odile Decq, l'AUC, Christian de Portzamparc, FAA+XDGA, Julien de Smedt, Annette Gigon and Mike Guyer, ANMA, François Leclercq and Marc Mimram, Hondelatte Laporte, Habiter Autrement, Stéphane Maupin, Brenac & Gonzalez, Thierry Beaulieu, adlib architecture
景观园林设计：Michel Desvigne
客户：SAS ParisNordEst, a company created by SEMAVIP (part-owned by the City of Paris), Caisse des Dpts and Icade
用地面积：5.5ha / 总建筑面积：165,000m² / 施工方：VINCI Construction France / 造价：EUR 1 billion (estimated)
竣工时间：2015.11 (building), 2016.4 (exterior spaces) / 摄影师：courtesy of A H A

In 2007, Rem Koolhaas and Floris Alkemade of OMA won the competition to masterplan Entrepôt Macdonald. Alkemade developed the plan's guidelines: to respect the existing monumentality and superstructure; preserve the warehouse's linearity but pass a tramline through its center; build new structures on the roof; introduce natural light; create mixed usage; and keep the original first-floor concrete facade of the 617 m north facade. A study confirmed that the concrete frame and its 8 x 8.5 m column grid were robust enough to support new development, including structures on the roof. Re-using the old building would preserve embedded energy and prevent pollution escape. The masterplan allowed 165,000 m² of mixed use (or 220,000 m² including car parking). 15 architects were appointed in 2008, each designing for different sections of the site, and conforming to crystal (clear) and mineral (opaque) horizontal stratification requirements along the facades.

This Entrepôt Macdonald includes 1,130 residences (of which 641 are social housing) over 74,300 m², a 16,000 m² business incubator, 28,000 m² of office and ground floor shops. A college, primary school, sports center, nursery and community center cover 16,300 m². One floor above the ground, the warehouse is split into two parallel parallelepids, separated by gardens between and a series of courtyards. A 18.5-meter-wide north side esplanade includes slopes down to road level, and the south faces the Rosa Parks station.

Infrastructure works began in 2010, and work on the structure in 2012. The tram service began in December 2012. Offices were completed and the college opened in 2014. The final building section, the incubator CARGO, was finished in November 2015. Shops opened and public spaces were finished in early 2016. Entrepôt Macdonald is now home to 3,500 residents, 4,000 workers and 1,000 students.

N4大楼视图，设计者：Julien de Smedt Architects view of the building N4, Julien de Smedt Architects

N2大楼视图,设计者:ANMA (Agence Nicolas Michelin & Associés) view of the building N2, ANMA (Agence Nicolas Michelin & Associés)

仓库中央花园视图和S2大楼前景视图,设计者:Brenac & Gonzalez view of the garden at the heart of the Entrepôt and in the foreground view on the building S2, Brenac &Gonzalez

S7大楼内部视图,设计者: Hondelatte Laporte Architectes
interior view of the building S7, Hondelatte Laporte Architectes

企业孵化园北立面视图,设计者: Studio Odile Decq
办公室,设计者: François Leclercq & Associés + Marc Mimram Architect
north facade, view of the incubator, Studio Odile Decq
offices, François Leclercq & Associés + Marc Mimram Architect

左侧N4大楼,设计者: Julien de Smedt/右侧连桥建筑和S6、N5大楼,设计者: FAA + XDGA building N4 on the left, Julien de Smedt / bridge-building / S6 and N5 on the right, FAA + XDGA

英国设计博物馆
OMA + Allies and Morrison + John Pawson

英国设计博物馆的新家位于伦敦西部肯辛顿主街,改造后于2016年11月24日对外开放。该馆位于一栋建于20世纪60年代的标志性现代主义二级列管建筑中,该项目历经了五年的施工过程。相比位于伦敦东南部西鲱泰晤士的旧博物馆建筑,当前博物馆的面积增加了两倍,达到10000m²。

OMA、Allies & Morrison和Arup保留了建筑壮观的混凝土屋顶和独特的立面。通过重塑建筑内部,约翰·帕森在栽种了一排橡树的中庭周围营造了一系列平静并具有美感的空间,并结合了建筑原始结构中的关键要素。

他们进行了大量复杂的翻新工程,包括结构整体重组和地下室的挖掘工作,以增加其建筑面积和组织效率,以此来满足英国设计博物馆的需求,同时,在取得了遗产保护官员的同意之后,保留了引人注目的屋顶底面结构。在实现了翻新工程的同时也保留了著名的抛物线形式的铜屋顶,而这离不开Arup和承包商Mace的重要工程技术。

建筑师将建筑外立面完全替换掉,以符合当代的建筑技术标准;对玻璃立面进行了重新设计和替换,以保留先前的开窗模式和RHWL大楼原始的蓝色玻璃外观。这一新的玻璃系统可以控制入射日光,也能让人欣赏到未来博物馆外部的景色。原来的彩色玻璃面板被拆下来翻新和恢复原状,以供未来前来博物馆参观的人们欣赏。

进入博物馆,参观者可以在室内中庭中仰望到标志性的双曲抛物面屋顶的壮观景色。这尺度惊人的混凝土屋顶跨越整个建筑的长度,从建筑两个相对的拐角位置高高耸起,在上方形成了一个蝠鲼状的结构。主中庭的周围设置有画廊、学习空间、咖啡馆、活动空间和商店,好似一个露天矿区,参观者可以轻松地在这个空间中穿行,并且只要沿橡木楼梯上楼就可以看到建筑的全貌。

该联邦协会翻新项目已大功告成,这是Chelsfield LLP和llchester Estates公司开发的相邻的荷兰绿色项目中的一个重要组成部分。三个立方体石块状建筑的布置方式引人注目,并与保留下来的博物馆建筑的几何形状和网格相互呼应,同时在灵敏度极高的城市/公园环境中提供了54套住宅公寓。

The Design Museum

On 24 November 2016, the Design Museum opened in its new home on Kensington High Street, west London. Housed in a landmark grade II listed modernist building from the 1960s, the project is the culmination of a five-year construction process. The museum has now tripled to 10,000 sqm from its previous premise in Shad Thames, south-east London. OMA, Allies and Morrison and Arup have restored the building's spectacular concrete roof and distinctive facade. Remodeling the interior, John Pawson has created a series of calm, atmospheric spaces ordered around an oak-lined atrium, incorporating key elements from the original structure.

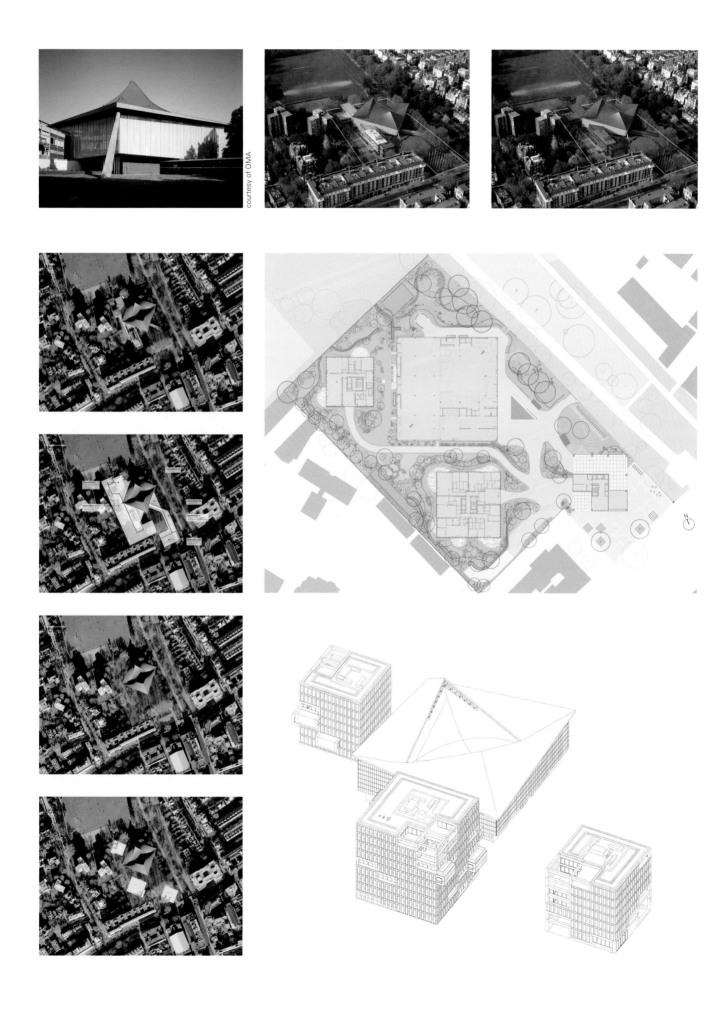

西立面 west elevation

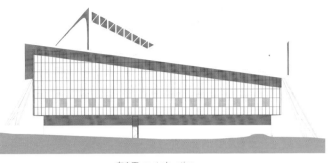
东立面 east elevation

Significant and complex refurbishment works were carried out, including the wholesale reconfiguration of the structure and basement excavation to increase the floor area and organisational efficiency to suit the needs of the Design Museum, while balancing the retention of the dramatic views to the underside as agreed with heritage officers. The refurbishment was realised while retaining the renowned parabolic copper roof in-situ, which required significant engineering skill from Arup and the contractor, Mace.

The facades have been completely replaced to fulfil contemporary technical building standards. The glazing was redesigned and replaced to retain the pattern of the fenestration and the blue-glass appearance of the original RHWL building. This new system permits controlled daylight into and views out of future museum spaces. Original stained glass panels were removed, refurbished and reinstated to be enjoyed by future visitors to the museum.

Inside the museum, visitors find themselves in a central atrium with striking views up to the iconic hyperbolic paraboloid roof. The stunning concrete roof spans the length of the building, rising on the two opposing corners to create a manta ray-like structure above. Galleries, learning spaces, café, events space and a shop are arranged like an opencast mine around the main atrium, allowing visitors to navigate the space with ease and to discover everything the building has by simply walking up its oak staircases.

The Commonwealth Institute refurbishment project has been realised as an essential part of the adjacent Holland Green development by Chelsfield LLP and Ilchester Estates, a striking arrangement of three stone cubes that respond to the geometry and grid of the retained museum building, providing 54 residential apartments placed within a highly sensitive urban/park context.

项目名称：The Design Museum / 地点：Kensington High Street, Holland Park, London, UK
博物馆设计团队：John Pawson, Arup, ChapmanBDSP, Turner & Townsend, Gardiner & Theobald LLP, Tricon Limited / 标识设计：Cartlidge Levene / 餐厅设计：Universal Design Studio, Prescott & Conran / 永久性装置设计：Studio Myerscough / 开发商：Chelsfield Developments Ltd in collaboration with the Ilchester Estates
开发商设计团队：OMA(Partner in charge: Reinier de Graaf), Allies and Morrison, West 8, Arup and AECOM / 开发商的承包方：MACE / 装备承包方：Willmott Dixon Interiors / 展览装置设计：Elmwood / 楼面和橡木面板：Dinesen / 零售空间设计：Vitra / 酒吧设计：Benchmark / 家具设计：Vitra / 照明设计：Concord / 建筑面积：10,219.33m² / 总建筑面积：2,694.19m² / 设计开始时间：2008 设计竣工时间：2016 / 摄影师：©Luke Hayes (courtesy of OMA) - p.100~101, p.105; ©Philip Vile (courtesy of OMA) - p.107, p.108, p.109top; ©Gareth Gardner (courtesy of the Design Museum) - p.106top, p.110, p.112~113; except as noted

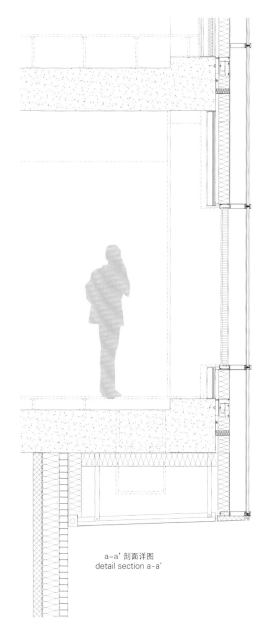

a-a' 剖面详图
detail section a-a'

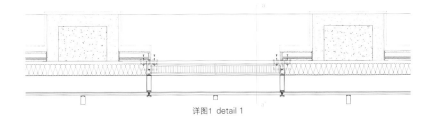

详图1 detail 1

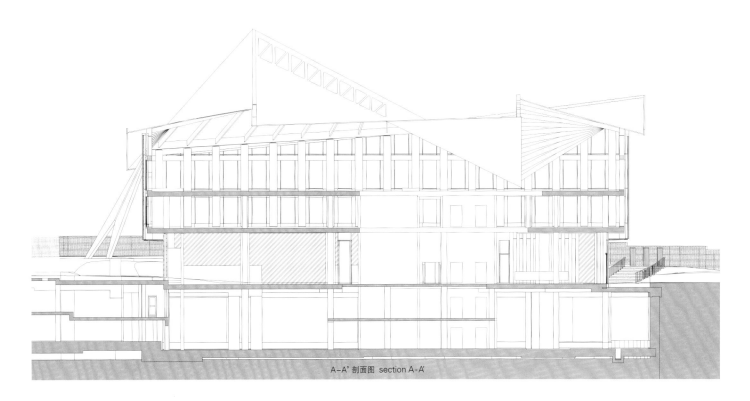

A-A' 剖面图 section A-A'

©Nick Guttridge (courtesy of OMA)

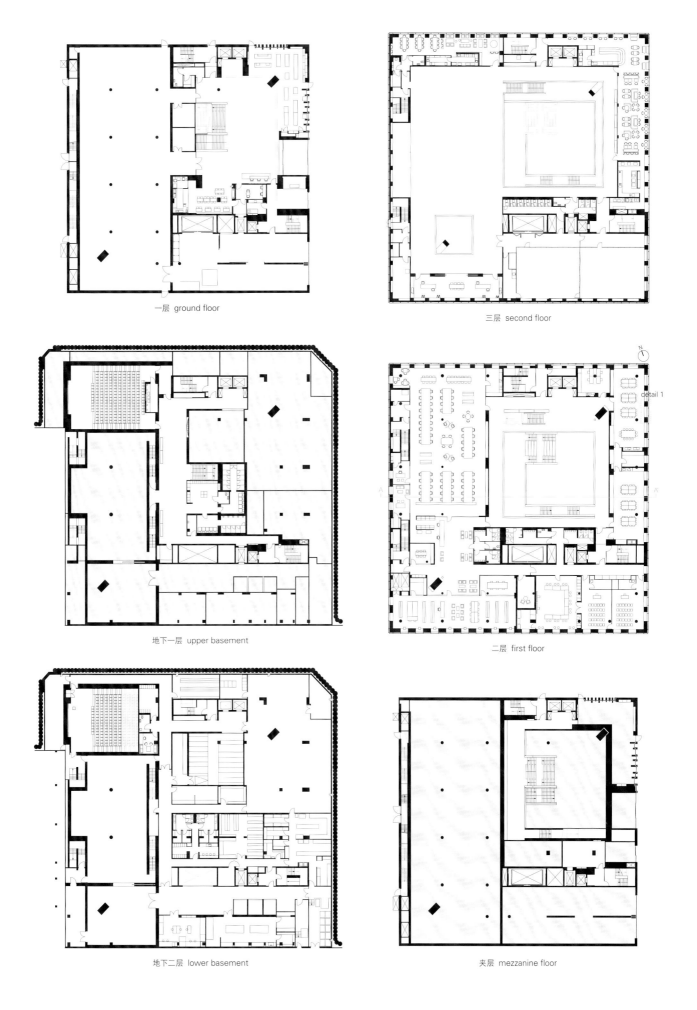

一层 ground floor

三层 second floor

地下一层 upper basement

二层 first floor

地下二层 lower basement

夹层 mezzanine floor

马德雷拉古堡
Carquero Arquitectura

当时光跨越屋顶 When Time Jumps through the Roof

因地处最新的"那扎里"边界——瓜达莱特河谷与贝提卡山脉的交汇处这样的战略性地理位置，这座中世纪的塔楼一直以来都是具有历史意义的地标。然而2013年，部分塔楼倒塌，失去了部分壮观的体量，不仅导致塔楼的其他部分存在失去稳定性的危险，同时也使建筑可能失去其作为关系到当地形象和文化的景观标志物的地位。因此，我们这个项目的目的在于对这一景观标志物进行加固。

因为经济危机，许多城市和地区的文化形态和标志都遭受了巨大的损失，由于对这些老建筑的维护并不能带来经济利益，所以它们日渐衰落。

根据相容性和真实性标准，本次建筑干预着眼于对濒临倒塌的结构进行加固，以区分扩建部分和原始结构，避免出现我们法律禁止的模拟重建行为，并恢复塔楼作为景观标志物原有的体量和色调。

因此，本项目的重点并不是想要呈现一个未来的形象，而是反映建筑的过去及其自身的历史起源。该项目着眼于统一的潜在恢复，而

并未承担建造伪造的历史名胜的任务或消除随着时光流逝留下的每一寸痕迹。本项目试图通过承认建筑遗迹的物理一致性和双极性（美学和历史方面）中的"纪念性"（回忆）来开展修复工作，以便将这两个方面传递给未来。

所有早先进行的历史、构造、功能、结构和倒塌原因分析以及考古监督都对确定项目的设计细节至关重要。而施工过程中的考古新发现找回了有关这栋文化历史遗迹的部分有趣的历史。

倒塌的那些石灰岩被重新用于建造新建筑的扶壁，以确保其稳定性，并为失去外墙砌体结构而变矮之后的内部核心提供加固/保护。为加固倒塌后保留下来的、存在严重倒塌风险的狭长墙壁，建筑师选择将顶部去掉。在建筑外立面上，表面被移除，保留其内墙上原有的白色覆层以及一幅有趣的赭色小船的壁画。为显示出建筑原有的体量，建筑师还根据现有几何要素的细节重新考虑了建筑所有的边缘。

与对可移动的历史遗迹的实用性建筑干预一样，该建筑的历史价值同样得到了提升，建筑与原始体量相对，利用一个与原始覆层相似的连续涂层（石灰灰浆），不仅填补了缝隙，还使得人们可以解读这个嵌入式的建筑单元。同样，体量上部的外壳定义了几个不同的施工阶段，突出了原来隐藏于层层叠叠的地层之后的城垛射击口。

Matrera Castle

This medieval tower has always been a historical landmark due to its strategic position in the latest "Nazari" border, where the Guadalete Valley meets the Bética mountain range. However, it had partially collapsed in 2013, losing part of its imposing volume and putting at risk not only the architectural stability of the rest of the tower, but also its role as a landscape landmark, linked to the iconography and culture of the region. Therefore, our project was to consolidate such a landscape icon.

These cultural models and emblems of our cities and territories have suffered enormously due to the economic crisis, which has made them fall into decadence because of the lack of economic interest in their maintenance.

With compatibility and authenticity criteria, the intervention looks at structurally consolidating the elements that are at risk, to differentiate the additions from the original structure, avoiding mimetic reconstructions (that our law prohibits), and recovering the volume and tonality that the tower originally had as a landscape icon.

The essence of the project is not intended to be, therefore, an

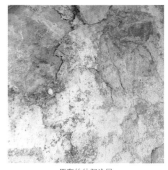
原有的外部涂层
original exterior coating

原有的内部涂层
original interior coating

干预之前的顶部细节
detail of the crown
before the intervention

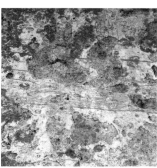
顶部原有的赭色壁画
original ochre fresco painting
in the crown

image of the future, but rather a reflection of its own past, its own origin. With brandian reference, this project aims to look at a unifying potential restoration, without undertaking the task of building a false historical monument or cancelling every trace of the passage of time. It tries to approach the work in recognition of the "monumentum" (memory) in its physical consistency and its dual polarity, aesthetic and historical, in order to transmit those two aspects to the future.

All that previous historical, constructive, functional, structural and pathological analyses, as well as the archaeological supervision that has taken place, have been important to define the details of the project. New discoveries have appeared during the works that have served to recover part of the interesting history of this monument of cultural interest.

The very limestone that had collapsed was re-used for the buttresses that guaranteed its stability and for the reinforcements/protections of the internal degraded cores that had lost their exterior stonework. The top was executed in order to consolidate the slender wall that remained after the collapse and that was running a serious risk of overturning. In its exterior face, the flesh was removed and the original white covering was retained in its interior face as well as an interesting fresco painting of a boat in ochre tonality. All the edges were also rethought from the details of existing geometrical elements, in order to mark its original volume.

In parallel to the practical intervention of movable heritage, its historical value has also been enhanced, facing its original volume using a continuous coating (mortar of lime) similar to the one which originally covered it, which clogs the gap and allows reading of the architecturally recessed unit. Likewise the upper casing defines its construction phases, enhancing the original battlements shots that were hidden behind their stratigraphic superposition.

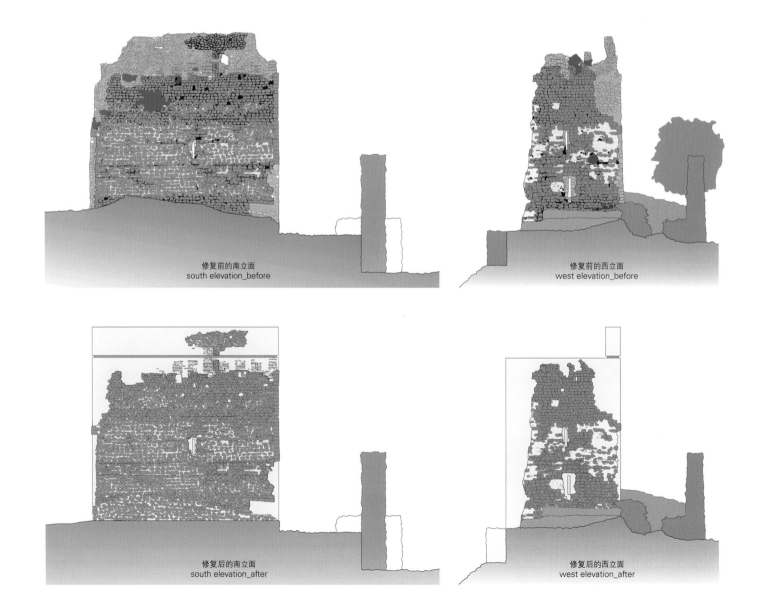

修复前的南立面
south elevation_before

修复前的西立面
west elevation_before

修复后的南立面
south elevation_after

修复后的西立面
west elevation_after

项目名称：Matrera Castle
地点：Villamartín, Cádiz, Spain
建筑师：Carlos Quevedo Rojas_Carquero Arquitectura
项目团队：Cristina Pérez Prado, Emilio García Chacón,
Joaquín Martín Rizo
技术建筑师：José Antonio Cabeza Pérez
考古学家：José María Gutiérrez
承包商：Arcobeltia Construcciones S.L.
客户：Ubri-Prado S.L. / 用途：restoration
用地面积：15,568m² / 建筑面积：136m²
建筑覆盖率：0.87% / 总楼面覆盖率：1.74%
结构：preexisting limestone
外墙饰面：lime mortar
设计时间：2012—2013 / 施工时间：2014—2015
摄影师：©Mariano Copete Franco +
Francisco Chacón Martínez (courtesy of the architect)

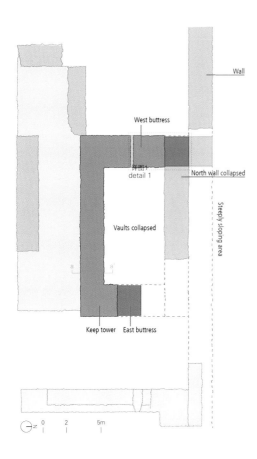

1. Reinforcing bars up to coronation
2. Mortar
3. Internal wall structure
4. Previous cap of lime mortar after cleaning
5. Core realized with the collapsed limestones
6. One feet brick wall
7. Lime mortar 15 mm
8. Fiberglass mesh
9. White rough lime mortar 15 mm

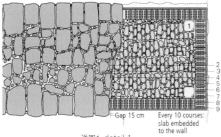

详图1 detail 1

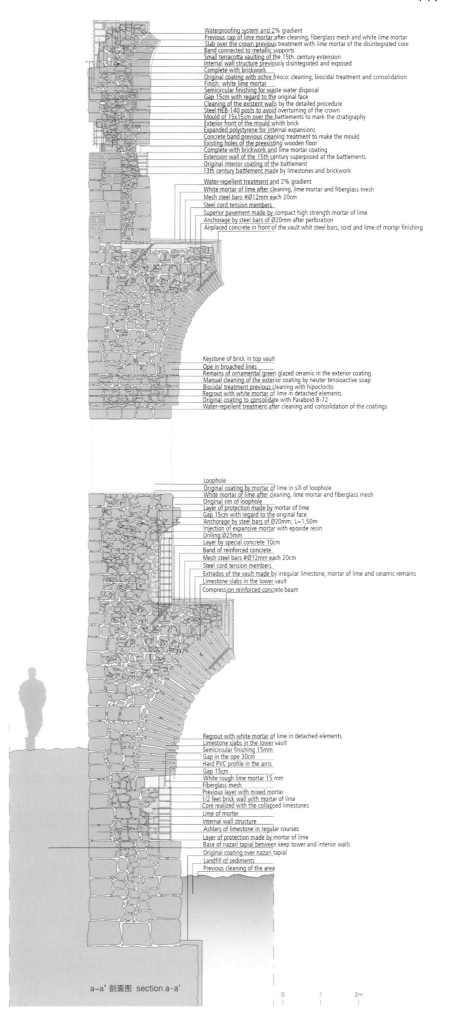

a-a' 剖面图 section a-a'

城市地标建筑新格局

New Configurations for Urban Landmarks

马尔默Live大楼_Malmö Live / Schmidt Hammer Lassen Architects
鹿特丹Timmerhuis大楼_Timmerhuis, Rotterdam / OMA

城市地标建筑新格局_New Configurations for Urban Landmarks / Silvio Carta

随着公共活动需要建筑之间的关联度日益增加，越来越多的城市变得越来越复杂，建筑设计也就越来越需要与反映这种复杂性的解决方案相呼应。我们可以从数字化原理的角度来解释这种情况，在该原理中，各个元素都被认为是模块化实体的有组织的系统。从更加严格的数字观念来看，基本实体将是输入/输出的数字，能将所有一切都转化为一连串的二进制信息。如果我们将它应用到建筑设计当中，那么与数字化原理相似，一个实体体量也可以被分解为一系列的模块化元素。

复杂的大规模项目通常要求在同一个巨大体量内实现明确的设计表达。在《癫狂的纽约》(1978年) 一书中，雷姆·库哈斯将摩天大楼的性质定义为在外观坚固而统一的外围护结构内组织不同活动的容器。在某种程度上，OMA的摩天大楼项目鹿特丹大厦 (2013年) 还体现了垂直城市的理念，在此，(空间、功能和活动的) 复杂性都被包裹在同一个外壳之中。由于醒目的外形和巨大的尺寸，这些巨大的建筑综合体变成了城市的地标建筑，并且它们的形象往往都得到了全世界的认同。然而，根据不断增加的使用需求，越来越多的新建筑必须包含各种超出其物理界限的活动。其设计中包含的功能与该城市的 (来自其他建筑、街道和广场以及这些场所内部形成的) 都市生活息息相关。但这一切都是如何发生的呢？

The more cities become complex, with the public activities requiring an increasing degree of interrelationship with buildings, the more architecture needs to respond with solutions that reflect this complexity. One way to interpret this scenario could be through the lens of the digital axiom, where elements are considered as organised systems of modular entities. In a more strict digital sense, the basic entities would be the I/O digits that translate everything into a series of binary information. If we apply this to architecture, where a solid volume can be broken down into a series of modular elements, one may appreciate the parallel with the principle of digitalisation.

Complex and large-scale programmes often require being articulated within the same massive volume. In *Delirious New York* (1978), Rem Koolhaas examined the nature of skyscrapers as containers of different activities within the same envelope, with a solid and uniform appearance. To a certain extent, OMA's skyscraper project De Rotterdam (2013) still embodies the idea of the vertical city, where complexity (of spaces, programmes and activities) is wrapped in the same shell. For their shape and size, these massive complexes become urban landmarks, and their image is recognisable often at the global scale. However, increasingly new buildings are required to incorporate a variety of activities that exceed their physical boundaries. The functions they are designed to host are entangled with the urban life of the city that flows from and within other buildings, streets and squares. But how does this all happen?

Complexity can be digitalised, namely rendered as a sequence of modular elements that are then combined systematically. In this case, space is no longer contained within the same block, but divided into its elementary modules. These

鹿特丹大厦，OMA，荷兰，2013年
De Rotterdam by OMA, The Netherlands, 2013

可以将复杂性数字化，即将其呈现为一系列模块化元素，然后将其系统化地结合到一起。在这种情况下，空间不再被包含于同一个体块当中，而是被划分为基本模块。随后，这些单元可以在空间中得到重新排列，从而在建筑内部和外部及其模块中间都可以举行不同的活动。大型建筑物作为地标建筑，需要延展城市功能，不仅在视觉上作为城市集体形象的一部分，而且更重要的是，发挥都市生活催化剂的作用。从这种角度看，人们就能够理解马尔默Live大楼和位于鹿特丹的Timmerhuis大楼的设计了。这两栋大楼都包含多种功能并且作为城市中枢发挥作用，可提供从住宅、酒店到公共和文化项目等一系列综合服务。这两栋综合建筑的设计目的均在于吸引访客，并产生高水平的公众参与度。从大楼前面和运河边的人行漫步道，到建筑内部的几处画廊，以及可以举行小型音乐会的屋顶空间，马尔默Live大楼在室内和室外提供了一系列方便人们相聚的公共空间。Timmerhuis大楼就位于鹿特丹市中心最受欢迎的街道之一——Meent大街的后方。一系列小型绿色区域、广场和运河环绕着建筑，在这里人们可以继续漫步在建筑内部的鹿特丹博物馆里，或在咖啡厅小憩。

OMA和Schmidt Hammer Lassen建筑师事务所以不同的方式在其设计的建筑中明确地表达了方案的复杂性，为人们提供了两个范例，展示他们是如何实现数字化体量的合理安排的。马尔默Live大楼由一系列体量组成，包含一个音乐厅、若干会议设施和一家酒店。根据周围建筑视线的不同，这些体量的高度和朝向各不相同。凭借其完全开放和自由进出的一层，以及与所处环境相呼应的几何形体，这栋大楼

units can then be spatially rearranged to allow for different activities to happen both inside and outside of the building shell, as well as in-between its modules. Large-scale buildings acquire an extended urban role as landmarks that not only function visually as part of the collective imagery, but also and more importantly, act as a catalyst for urban life. Within this perspective, the Malmö Live complex and the Timmerhuis complex in Rotterdam all can be read. The two buildings house multiple functions and operate as urban hubs, offering an integrated series of activities from residential and hotel to public and cultural programmes. Both are complex buildings designed to attract visitors and generate a high level of public engagement. Malmö Live provides a series of public spaces, both indoor and outside for people to gather, ranging from the promenade in front of the complex and facing the canal, to the several galleries inside the buildings and the roof tops spaces for small concerts. The Timmerhuis sits right behind Meent, one of the most popular streets in Rotterdam city center. A series of small green pockets, squares and canals surround the complex, where people can continue their stroll inside the museum Rotterdam or the café inside.

The way OMA and Schmidt Hammer Lassen Architects have articulated the complexity of the programme in their building varies, offering two examples of how the digitalisation of masses can be organised. Malmö Live consists of a series of volumes that host a concert hall, congress facilities, and a hotel. These have different heights and orientations, following sightlines of the surrounding buildings. With its totally open and free-to-walk-in ground floor, and geometries that relate to the context, this complex appears to be almost sculpted on site as a whole and subsequently fragmented into smaller

马尔默Live大楼,瑞典
Malmö Live, Sweden

的整体造型看起来几乎像是在场地上一气呵成的,随后再被分散成更小的部分。所有的体量都有一个相似的立面,它们在相同的主题之下设计有不同的图案和颜色。大楼的几个部分彼此之间看起来各不相同,但却是整体中密不可分的一部分。另一方面,Timmerhuis大楼的设计构思是由一系列模块化空间碎片构成建筑整体。许多空间单元格彼此堆叠在一起,然后,被小心地移位以调整从关键视角望去的视觉效果,并巧妙地与邻近的Stadstimmerhuis大楼——一栋1953年建造的市政大楼相呼应,而目前它成了新大楼的配套建筑。在该项目中,数字化组织甚至体现在结构布局上——使用7.2m×7.2m×3.6m的钢结构立方体模块形成了大楼的框架。该系统使大楼的首层实现了灵活性,并有了若干大型开放区域,而公共空间可以由内部街道向建筑的首层延续。从外部看,对于规划的前景,Timmerhuis大楼并没有显示出任何变化。上部的居住单元采用与中间楼层的办公室以及其他功能区相同的立面,为大楼提供了一个统一却又富有变化的整体外观。

这两个项目可被视为城市活动的变化方式的强有力的证据,城市活动正在变得越来越复杂、相互协调和舒适。同样,建筑也在不断地演化发展,包含新的外形和格局,在演化过程中巨大而坚固的建筑综合体被分解为较小的单元,而公共空间也日益受到欢迎,自如地融入到建筑当中。Timmerhuis大楼和马尔默Live大楼代表了全新一代的城市地标建筑,在这两个项目中,建筑综合体可以建立与城市的新关系,促进城市生活的繁荣。

parts. All the volumes have a similar façade which varies in pattern and colour within a common theme. The several parts of the complex appear distinct to each other, yet cohesively part of a whole. On the other hand, the Timmerhuis has been conceived as the result of a serial fragmentation of space in modular units. A number of spatial cells have been stacked on top of each other, then carefully shifted to adjust the visual impact from key viewpoints, and to subtly match with the adjacent Stadstimmerhuis, a 1953 municipal building now annexed to the complex. The digital organisation is here reflected even in the structural layout, where a 7.2x7.2x3.6 cubic meters steel structure cuboidal module is used to create the skeleton of the complex. This system allows for flexibility and large open areas in the ground floor, where the public space flows in from the street inside. From the outside, the Timmerhuis does not show any variation with regard to the outlook of the programme. The residential cells on the top are clad with the same façade of the offices in the middle levels, as well as the rest of the functions, providing the complex with a consistent, yet varied overall appearance.

The two projects can be regarded as a strong testimony to the way urban activities are changing, becoming increasingly more elaborated, integrated and nestled. By the same token, architecture is evolving, incorporating new shapes and configurations, where massive and solid complexes are broken down into smaller units, and the public space is growingly welcome to freely flow into them. Timmerhuis and Malmö Live epitomise a new generation of urban landmarks where complex buildings are able to establish new relationships with the city for urban life to flourish. Silvio Carta

马尔默Live大楼

Schmidt Hammer Lassen Architects

位于瑞典马尔默的新文化中心——马尔默Live大楼,是一座面积为54000m²的音乐会、会议和酒店综合设施。总体规划还包括面积为27000m²的住宅和商用空间。本项目将三种独立的建筑类型:酒店、会议和音乐厅结合进一座多元化的建筑当中,使建筑适用于不同的功能和活动。

马尔默Live大楼是一座开放的、富有表现力的动感建筑,其内部空间能够为各种活动提供场所。建筑设计采用现代北欧建筑的传统设计方法,注重清晰的功能型组织,还有易于通行的开放式底层平面布局。项目建成后成为马尔默市的焦点和地标式建筑,提供了一个展现城市精神、多样性和亲切感的环境。设计想法是创建一个"城市之家",结合马尔默市的城市环境开发一座建筑,对当前的城市生活做出贡献。

主入口位于建筑北面,是一个典型的长廊形状,对面是建筑前面的广场。在南面,游客可以从运河边的散步道直接进入建筑。

建筑的一层完全开放,并可以通往所有功能区,不论你是去音乐会、参加会议、去酒店住宿,或者只是小酌一杯咖啡,或者想要抄捷径穿过大楼,都能实现。多样化的空间功能组合在一起就像一座小城市。在建筑内部有三个体量,分别为大型交响乐大厅、多功能大厅和会议厅,通过这些定义明确的元素之间的组合,奠定了建筑的基调——音乐厅。建筑外部的街头生活被直接引入内部,以支持建筑开放的特性。

可以容纳1600个座位的音乐厅是马尔默Live综合大楼的核心。音乐厅长45m,宽23m,高20m,专门为马尔默交响乐团设计。音乐厅由橡木和黄铜材料建造而成,这些材料有助于形成一种亲密的气氛,并与综合体内部其他那些大胆采用未经处理的黑色混凝土和石料的室内设计形成对比。

后勤&功能安排
logistics & program arrangement

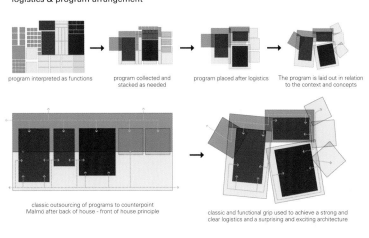

现场关系
on-site relationships

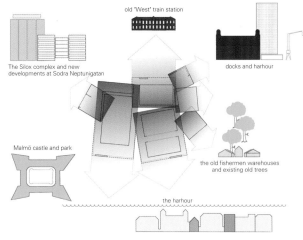

立面层次示意图
the facade layer cake

马尔默Live大楼的颜色
The colors of Malmö

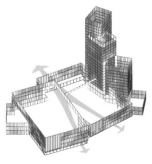

1. The Chinese lamps

The project is conceived as an open and democratic building where the ground floor is a public space. We wish to strengthen this effect by allowing the urban pavement continuing throughout the building and lifting the foyer level while at the same time opening the north/south ground floor facades.

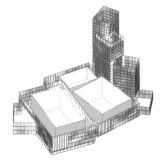

2. The Foyer cladding

The three main halls (Concert Hall / Black Box / Congress Center) are cladded with precast concrete elements (1200x8000mm), with three different depths, creating a vivid expression and breaking down the scale of these huge volumes

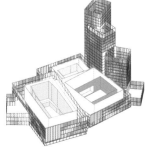

3. The Acoustic chambers

The innermost cores of the three halls are cladded with acoustic wood panels, according to each Hall's specific acoustic requirements. Each hall has however its own variation of the so-called Mondrian pattern.

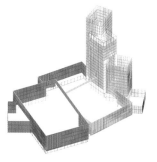

examples of color shades

concert hall   congress   hotel

马尔默Live大楼内部和外部的会议、音乐厅、集会设施以及公共空间使得该建筑能够适用于各种不同的活动。

建筑的多功能性及其在建筑内部和周边创造的生活氛围是企业长期发展能力的关键。酒店的加入使得这栋综合体全天候都在运作当中。在克拉里翁酒店&马尔默Live会议中心内部共有四家概念餐厅：一家空中酒吧高级餐厅、一家墨西哥餐厅、一家酒店餐厅和宴会厅以及一家熟食店和咖啡厅。

马尔默Live大楼将LEED白金等级视为目标，并且该建筑符合瑞士绿色建筑计划——"Miljöbyggprogram Syd"的A类标准。马尔默Live大楼是一个可持续的绿色项目，设计目标是以对环境影响最小的代价实现区域性开发。

Malmö Live

Malmö Live, the new cultural center in Malmö, Sweden is a 54,000m² concert, congress and hotel complex. The masterplan also includes 27,000m² for housing and commercial use. The project combines three separate typologies: hotel, congress center and concert hall into one multifaceted entity, enabling the building to adapt to many different functions and activities.

Malmö Live is an open, expressive, dynamic building offering numerous activities within its architecture. The point of departure for the building's design is the modern Scandinavian architectural tradition, which focuses on clear, functional organisation and an accessible, open ground floor layout. The building has become a focal point and landmark in Malmö, offering a setting in which the spirit of the city, its diversity and intimacy receive an architectonic expression. The idea was to create a "house of the city" incorporating the appearance of Malmö to develop a building that will contribute to the existing urban life.

The main entrance is found at the northern part of the building, which has a classic loggia motif facing the plaza in front.

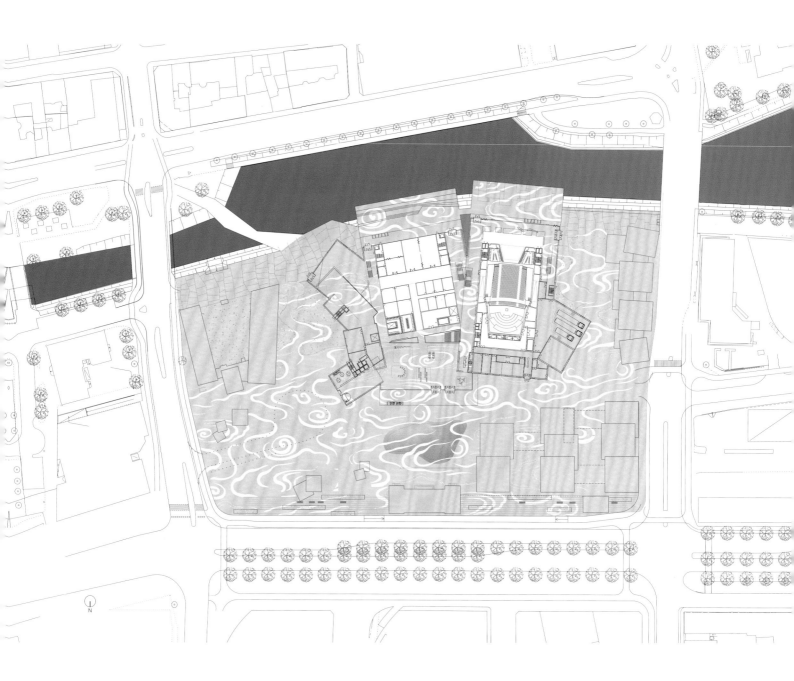

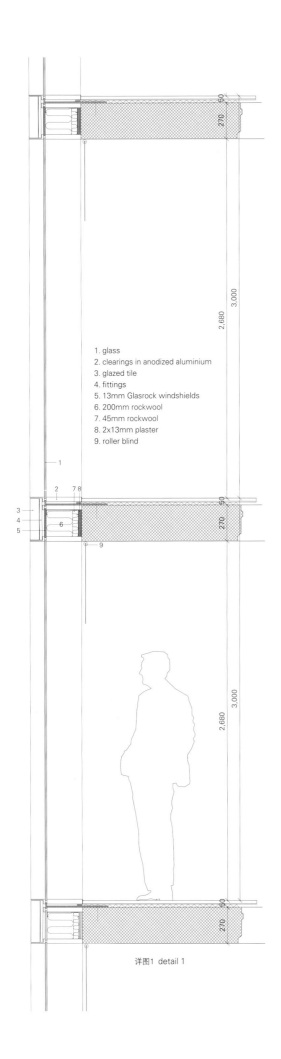

详图1 detail 1

1. glass
2. clearings in anodized aluminium
3. glazed tile
4. fittings
5. 13mm Glasrock windshields
6. 200mm rockwool
7. 45mm rockwool
8. 2x13mm plaster
9. roller blind

From the south, visitors enter the building directly from the promenade running along the canal.

The ground floor is fully open and accessible to all, whether going to a concert, attending a congress, staying at the hotel, or just stopping by for coffee or taking a shortcut through the building. The different functions are organised as separate elements to resemble a small city. Inside, three volumes hold a large symphony hall, a flexible hall and a conference hall, which are clearly defined elements that, through their mutual composition, set the tone for the musicality of the building. The street life outside is drawn directly inside to support the open nature of the building.

The 1600 seat concert hall is the center piece of the Malmö Live complex. Its 45m long, 23m wide and 20m high dimensions are purposely designed for the Malmö Symphony Orchestra. The concert hall is made of oak and brass which help create an intimate atmosphere and stand in contrast to the complex's other interior that stands boldly in raw black concrete and stone.

The flexibility of the congress, the concert hall, the meeting facilities and the public spaces inside and outside Malmö Live makes it suitable for all kinds of events.

The multifunctionality and the life which it creates inside and around the buildings are essential for long-term commercial viability. The integration of the hotel adds 24/7 life to the complex. Within the Clarion Hotel & Congress Malmö Live there are four concept restaurants: a Skybar with fine dining, a Mexican restaurant, a hotel restaurant and banquet and a deli café and coffee shop.

Malmö Live targets LEED Platinum and the building meets category A standard of the Swedish green construction programme "Miljöbyggprogram Syd". Malmö Live is a sustainable, green project designed with the intention to develop the area with as little environmental impact as possible.

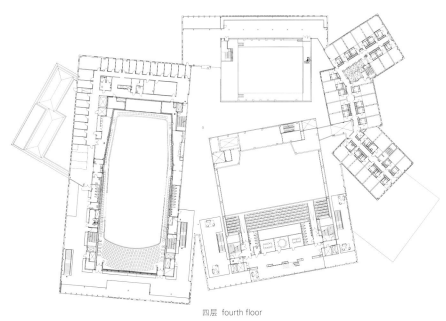

四层 fourth floor

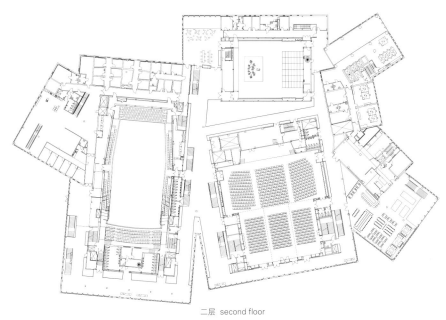

二层 second floor

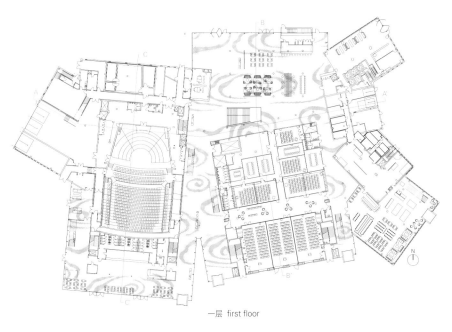

一层 first floor

项目名称：Malmö Live
地点：Malmö, Sweden
建筑师：Schmidt Hammer Lassen Architects
合作方：Akustikon
景观设计师：SLA
承包方：Skanska Sverige AB
客户：Concert Hall / City of Malmö / Hotel and congress / Skanska Sverige AB
可持续性设计：Tagetting LEED Platinum
用地面积：35,000m²
总建筑面积：43,000m²
竞赛时间：2010
竣工时间：2015
摄影师：©Adam Mørk (courtesy of the architect)

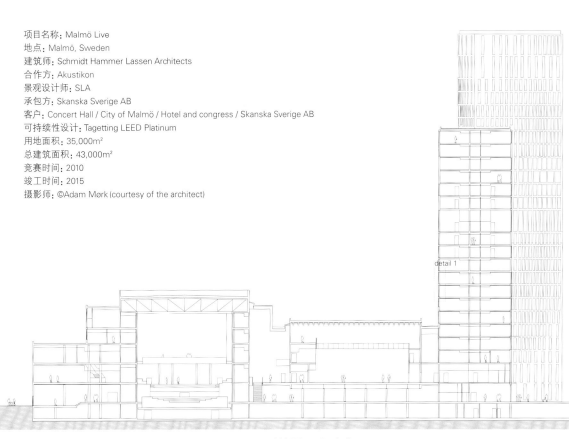

A-A' 剖面图 section A-A'

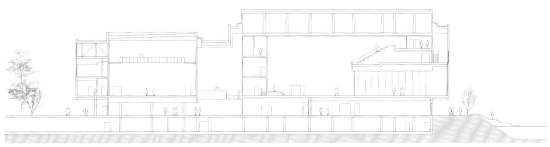

B-B' 剖面图 section B-B'

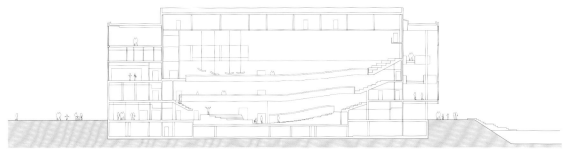

C-C' 剖面图 section C-C'

鹿特丹Timmerhuis大楼
OMA

城市地标建筑新格局 New Configurations for Urban Landmark

鹿特丹Timmerhuis多功能大楼为市政厅的新建大楼，将容纳市政服务设施、办公室和住宅单位。OMA的设计把Timmerhuis大楼构想成有着重复单元的模块式建筑，模块反复向上堆砌成两个不规则的塔尖，位置越高，模块就越往内退、离街道越远。Timmerhuis大楼由小型单元组成，从库尔辛格大街（鹿特丹一条主干道）看过去复杂而震撼；而此建筑物与Stadstimmerhuis（建于1953年的市政建筑）相邻的两侧，则显现出了细腻、适应能力强的一面。

Timmerhuis大楼的创新结构系统设计使建筑构造和功能布局同样高效而灵活：模块可按需要改装成办公空间或住宅。高层设有绿化阳台，可在鹿特丹市中心的城市环境中打造出附带花园的公寓。结构模块悬于街面之上而非入侵狭窄的城市空间，从而在地面层留出了广阔的开放空间，使Timmerhuis大楼和鹿特丹市之间能有积极主动而开放的互动。

设计大纲规定Timmerhuis大楼必须成为全荷兰最环保的建筑。为满足这项要求，OMA以灵活性作为大楼的核心概念，并设计了两个发挥"肺部"作用的大型中庭。两个中庭均连接气候系统，夏季储热而冬季蓄冷，再按需要释放冷气或暖气。大楼的三层玻璃幕墙立面应用高科技的半透明保温层，其能源效率之高史无前例。

鹿特丹过去的建筑设计好多采用修正主义规划，建筑风格毫不和谐；Timmerhuis大楼并不期望加入其中，而是试图以其模糊的体量成为周边现存建筑之间的缓冲。现有的市政厅和邮局之间的轴线与Timmerhuis大楼的对称轴重叠，而市政厅和邮局之间的街道也延伸成为通往Haagseveer大道的通道。Timmerhuis大楼与旁边的Stadtimmerhuis大楼保持相同的层高，两者因而协调相融，而基座也配合周边Laurenskwartier街区的特色而设计为20m高。

Timmerhuis, Rotterdam

For Rotterdam's Timmerhuis, a new building for the city hall that accommodates municipal services, offices and residential units, OMA conceived a modular building with repeated units gradually set back from the street as they rise into two irregular peaks. The building's composition of smaller cells creates an impressive, complex form when viewed from Coolsingel, one of Rotterdam's main arteries, and allows for subtlety and adaptability as the new building meets the Stadstimmerhuis (a municipal building, from 1953), which surrounds it on two sides.

The Timmerhuis's innovative structural system generates maximum efficiency and versatility both in construction and in program: units can adapt to either office space or residential parameters as desired. Green terraces on higher levels provide the possibility of an apartment with a garden in the heart of Rotterdam. On the street level, the structure allows

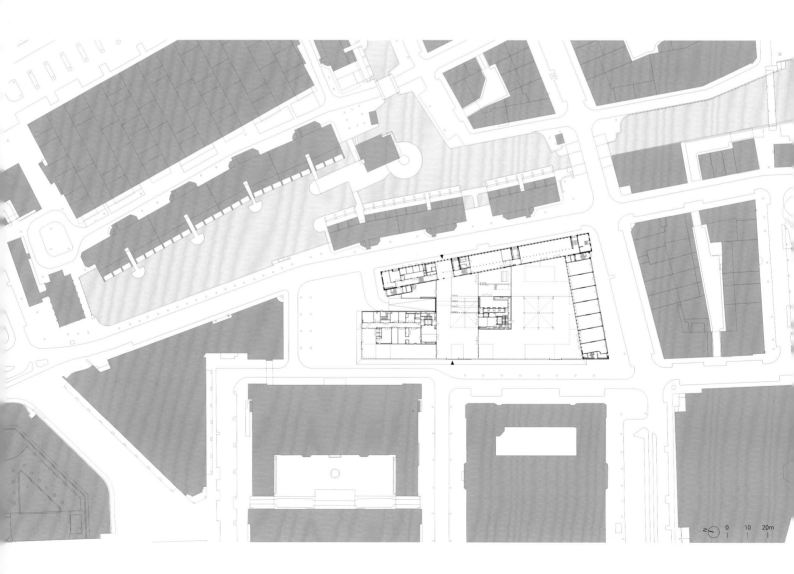

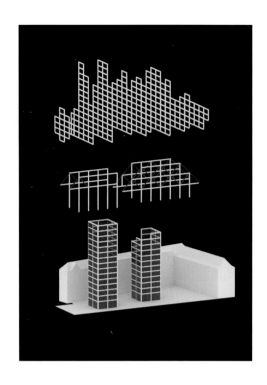
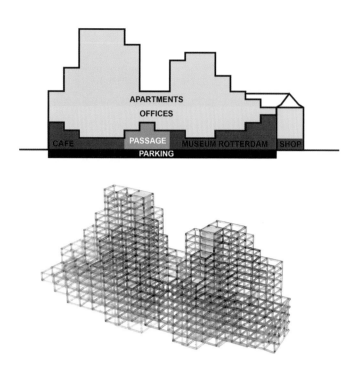
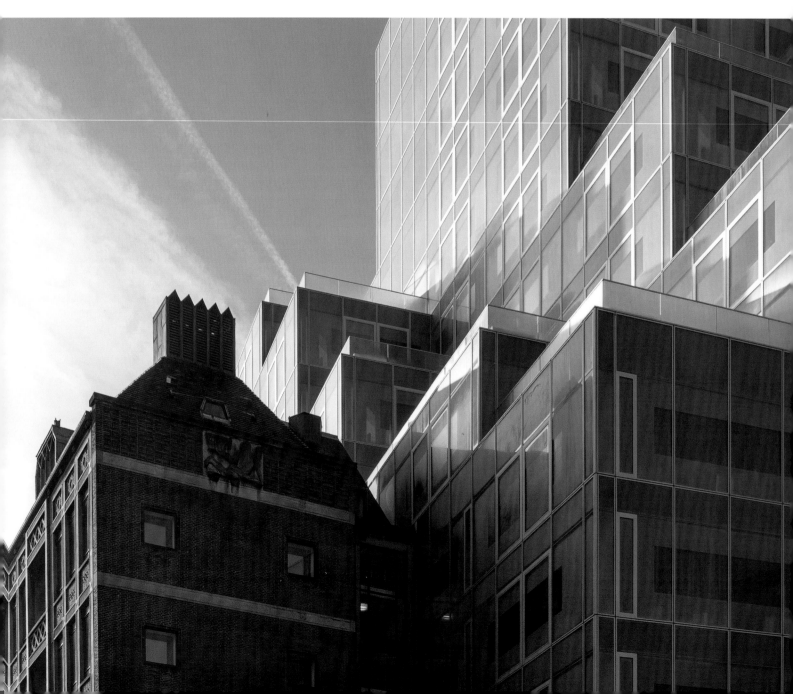

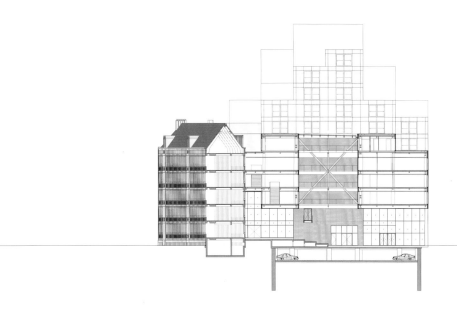

A-A' 剖面图
section A-A'

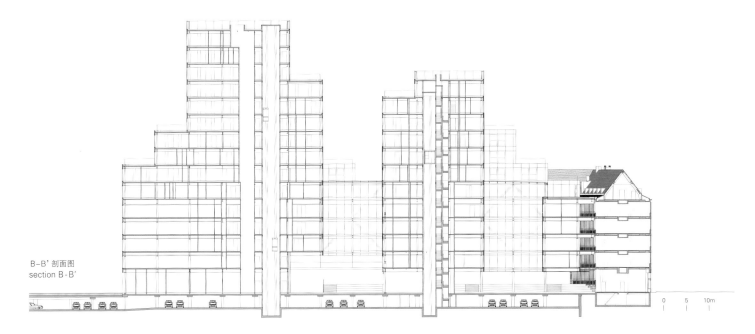

B-B' 剖面图
section B-B'

一层——办公室/住宅/博物馆/零售店
ground floor_office/housing/museum/retail

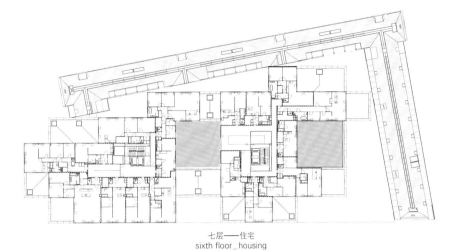

七层——住宅
sixth floor_housing

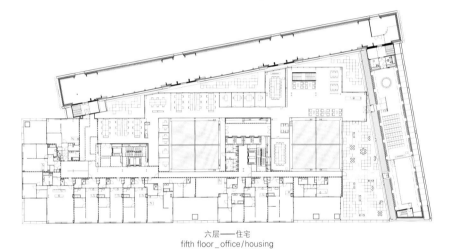

六层——住宅
fifth floor_office/housing

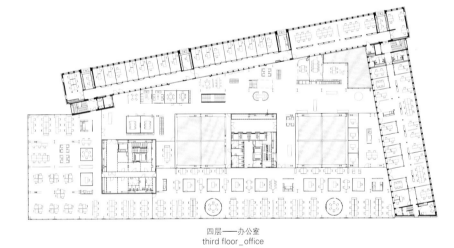

四层——办公室
third floor_office

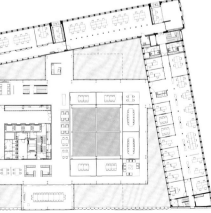

二层——办公室
first floor_office

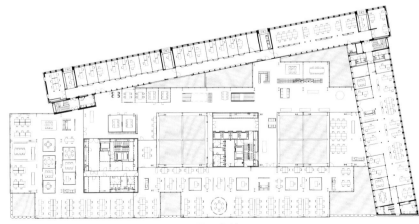

三层——办公室
second floor_office

for generous open space, with modules overhanging rather than encroaching into an interstitial area, encouraging an active and open engagement between the Timmerhuis and the city.

The design brief stipulated that the Timmerhuis must be the most sustainable building in the Netherlands. OMA tackled this imperative through the building's core concept of flexibility, and also through the two large atriums, which act like lungs. They are connected to a climate system that stores warmth in summer and cold in winter and releases this energy as warm or cold air as required. The building's triple glazed curtain wall facade uses hi-tech translucent insulation that allows for unprecedented energy efficiency.

Rather than being yet another statement in Rotterdam's crowded history of revisionist planning and cacophony of architectural styles, the ambiguous mass of the Timmerhuis tries to mediate between the existing buildings surrounding it. The axis between the existing town hall and the post office coincides with the axis of symmetry of the Timmerhuis, and the street between these two buildings continues into a passageway to the Haagseveer. The Timmerhuis integrates with the neighboring Stadtimmerhuis by maintaining the same floor heights, while the plinth height of 20m conforms to the character of the surrounding Laurenskwartier.

项目名称：Timmerhuis, Rotterdam / 建筑师：Rotterdam, The Netherlands
主管合伙人：OMA / 主管副合伙人：Reinier de Graaf / 竞赛阶段参与者：Alex de Jong, Katrien van Dijk
Competition phase: Reinier de Graaf, Rem Koolhaas, Partner in charge; Mark Veldman, Project leader / 概念设计/初步设计/扩初设计：Reinier de Graaf, Partner in charge; Alex de Jong, Associate in charge / 项目建筑师：Cock Peterse _ Construction documents; Saskia Simon, Katrien van Dijk _ Interiors / 工程、结构、装置，本地建筑师，造价顾问：ABT / 客户：Stadsontwikkeling Rotterdam
造价：approx. €100 million / 用地面积：5,000m² (including the old Stadstimmerhuis; opposite of Rotterdam City Hall) / 功能：office _ 25,400m², both existing and new part of the building; residential _ 12,000m², 84 apartments; exhibition space _ 1,630m²; retail _ 2,070m²; underground parking _ 3,900m², 120 places / 建筑高度：60m, 14 storeys in the North Tower and 11 storeys in the South Tower / 施工时间：2009—2015
摄影师：©Sebastian van Damme (courtesy of OMA) (except as noted)

海运集装箱的改造再利用

装载货物并将其运送至远方的简陋的海运集装箱,通常只会与货物和造船厂联系在一起,但长期以来,在当代建筑中它已成功被用作一种资源丰富的结构材料。鉴于模块化的外形,集装箱能够"像乐高积木一样"堆叠和组装。20世纪50年代,美国运输企业家马尔科姆·麦克林开始使用陆续实现标准化的运输集装箱,实现了交通基础设施之间的多式联运。最终,这种不起眼的集装箱通过创造性的建筑解决方案呈现出了一种全新的空间活力。

将这种现有的日常结构改造成创新型的居住空间具有很多革新优势,特别是在施工/建造阶段。标准化的尺寸和抗压钢构架是方便运输、设计和建造的关键特性。在进行模块化应用之前,先对结构进行切割、焊接并连接到其他集装箱上,或设计成一个组合,或设计成现有建筑和材料之间的寄生结构。然而,尽管为了实现材料的回收再利用,这些过程看起来是可取的,我们也必须牢记要考虑其中的环境因素。隐性因素和环境成本,如其中的能源消耗可能极高。当运送完货物时,如果不能装载回程货物的话,剩余的集装箱将不会全部被运输回去,因为如果运输空集装箱的话会造成浪费,所以造成了大量的堆积如山的废弃集装箱,数量令人担忧。

因此,如今,我们看到熟悉的海运集装箱被应用于各种各样的环境之中,有的在吉尔吉斯斯坦熙熙攘攘的市集,有的成为纽约布鲁克林的现代化、丰富多彩并且整洁的居住空间。集装箱已经成为生活基础设施的一部分。其简单的外形和模块化的组件为无缝堆叠、分层和模块化提供了方便。然而,在施工过程中需要重型机械、人工和切割设备。在施工准备阶段还需要对集装箱表面

Commonly associated with goods- and shipyards, the humble shipping container, loaded with cargo and destined for far-away places has been successfully used as a resourceful structural material in contemporary architecture for some time. Thanks to its modular form, the containers can be stacked and assembled "like blocks of Lego". In the 1950s, American transport entrepreneur Malcolm McLean started using what continued to develop into standardised transport containers, enabling intermodality between transport infrastructures. Eventually, this unpretentious container took on a new, spatial lease of life, through creative architectural solutions.

The remodeling of this existing everyday structure into innovative habitable spaces has many progressive advantages, particularly during the construction/build stages. Standard dimensions and stress tolerant steel framework are key features easing transportation, design and build. Preparing the structures for modular use, they are cut, welded and connected to other containers or designed as a combination or as parasitic architecture between existing buildings and materials. However, although the process appears to be a worthy option for recycling materials, we must remember that there are environmental factors to consider. Hidden factors and costs to the environment, such as energy use, can be extremely high. When the cargo is delivered, the surplus containers are not all shipped back again if they cannot be filled for the return journey, thus becoming uneconomical to ship back empty, creating mountains of disused containers stacking up in worrying quantities.

城市船舱_Urban Rigger / BIG
Ccasa青年旅店_Ccasa Hostel / TAK Architects
艾杰大学的高新科技园_Ege University Technopark / ATÖLYE Labs
锡安音乐中心_Sion Music Center / Savioz Fabrizzi Architectes
魔鬼之角_Devil's Corner / Cumulus Studio

海运集装箱的改造再利用_Re-use/Re-model – the Shipping Container / Heidi Saarinen

进行喷砂清理，使用重型机械将其吊运并放置于施工场地之上，造成较高的能量消耗。

　　Savioz Fabrizzi建筑师事务所设计的位于瑞士的锡安音乐中心，由当地的一个艺术计划提供资金赞助，该项目将原始状态的海运集装箱作为设计起点。该方案以其无所不在的古怪细节不断地给人带来惊奇，同时对材料中蕴含的原始却又吸引人的货物运输传统致敬。人们可以在吧台区发现原来集装箱钢板的下脚料，包括特有的扭锁把手。通过将"新"集装箱结构增加至现有的建筑设计上，一件令人满意的、未经加工的自然融合作品诞生了，在观众和表演者之间充满活力的互动中也清晰可见。在模块化框架之上增加或去掉集装箱的多种设计可能性给设计带来了更多的灵活性。搭配智能照明设备，材料赋予空间一种活泼的气氛。音乐中心以其充满趣味性的舞台表演及建筑内部其他地方的表演吸引了参观者的目光。

　　在魔鬼之角酒窖入口观景台可以欣赏到塔斯马尼亚东海岸的壮观景色，观景台将参观者直接引入大自然的中心。对再利用海运集装箱进行加工处理并露出外部的钢质立面，搭配木质覆层细节设计，并结合了采用崭新、光滑表面的同样整洁的室内。Cumulus Studio精心设计了观景点和室外区域，这里有专门供人会面、休息和沉思的露台。建筑的室内呈现出一种平静的气氛，回应周围大自然的广阔无垠，营造了一个吐纳和冥想的空间。

　　由BIG设计的创新型、碳平衡、富有活力的新式学生公寓使用了海运集装箱，在水面上下打造了干净的当代居住空间。当需求

Today, therefore, we see the familiar shipping container used in a range of settings from bustling bazaars in Kyrgyzstan to contemporary, colourful and clutter free living spaces in Brooklyn. The container has become part of our infrastructure. The ease of its form and modular components gives way for seamless stacking, layering and modularisation. However, the construction process involves heavy machinery, labour and cutting equipment. Preparation of the container also involves sandblasted cleaning of surfaces, lifting and positioning onto site, using heavy machinery, contributing to the high-energy usage.

Funded by a local arts scheme, Sion Music Center by Savioz Fabrizzi Architectes in Switzerland sees the shipping container in its original form, as a starting point to the design. The scheme continues to surprise with quirky details throughout; celebrating the raw yet attractive cargo heritage within the material. Canny off-cuts of the original container's steel panels, including typical twistlock handles can be found in the bar area. By adding the "new" container structure to existing architecture, a cool, raw fusion is created, also visible in the vibrant interaction between people and performers alike. Possibilities to add or subtract from the modular framework allow flexibility. Paired with clever lighting the material gives the space a lively ambience, seducing the visitors with its fittingly mischievous performances on stage and beyond within the industrial architecture.

With spectacular views over the Tasmanian East Coast, Devil's Corner cellar door and lookout brings visitors

马尔科姆·麦克林在铁轨边,纽瓦克港,1957年
Malcolm McLean at railing, Port Newark, 1957

　　上升时,还可以增加额外的空间。环保因素成为该方案的重点部分,比如,海运集装箱的运输和服务设备,还有供暖设备、太阳能装置和低能耗泵。在学生上完课之后,有什么比回到漂浮于哥本哈根的美丽海港中心超赞的"城市船舱"公寓更令人向往的呢?可惜我只能想象。

　　越南首家背包客旅店——Ccasa青年旅店由TAK建筑师事务所同样使用海运集装箱建造而成。接近项目场地的街景与旅店正面的钢质立面及随意的建筑和工艺材料形成了互动,巧妙地沿立面排成一条直线。建筑师将空间和设计的重点放在共享的公共空间上,鼓励来自全球的背包客进行互动。洗漱和洗衣设施也是共享的。采用明亮原色的海运集装箱相互堆叠,其内部为卧室,故意设计为小型空间,更具有"临时性",同样,这也是为了使建筑能有更多的空间促进旅行者在公共空间中的积极互动。

　　由ATÖLYE Labs设计的艾杰大学的高新科技园成为教育、前沿研究与行业合作伙伴之间跨领域合作的典范。考虑到当前大学校园场地的现状条件(包括一个被拆毁的建筑场地),建筑师起初就高瞻远瞩地决定将该项目的重点放在因地制宜考虑场地的特点上。设计团队通过使用回收再利用的海运集装箱为闲置的场地注入了生机。高楼层和低楼层通过走廊相结合,在每条小路上,都能看到光与影之间的自然邂逅和闪动。从高层欣赏到的景致和街景不仅让用户得到了视觉享受,同时还在更广阔的校园景

straight into the heart of nature. Reused shipping containers, treated and left to expose the steel facade on the exterior together with timber cladding detailing, combine the equally clean interiors in new, sleek surfaces. Cumulus Studio carefully choreographed viewing points and outdoor areas, with terraces specifically placed for meeting, rest and reflect. There is a calm within the building, echoing the surrounding vastness of nature, a space for breathing and contemplation.

An innovative, carbon neutral young and buoyant alternative to student accommodation, designed by BIG uses shipping containers to create clean, contemporary spaces on levels above and below water. Additional spaces can be added when demand rises. Environmental factors are heavily part of the scheme, from the transportation of the shipping containers through to the services: heating, solar power and low energy pumps. What more would a student want to go back to after lectures than the cool Urban Rigger floating pad in the center of Copenhagen's beautiful harbour? I can only imagine.

Vietnam's first back packers' hostel, Ccasa Hostel built by TAK Architects also uses shipping containers. On approach to the site, the street scene interacts with the front facade of steel and haphazard building and craft materials, cleverly aligned along the elevation. Spatial and design focus is on the shared, communal spaces, encouraging interaction between the global back packers. Wash- and laundry facilities are also shared. Stacked shipping containers in bright primary colours house the bedrooms, deliberately made smaller, more "temporary", again, allowing more space for positive engagement between the travelers whilst in the common areas.

The Ege University Techno-park by ATÖLYE Labs accommodates a role model for interdisciplinary collaboration between education, cutting edge research and industry partners. The architects decided early on that the project

集装箱运货船，法国达飞海运集团的克里斯托弗·哥伦布号货船，德国韦德尔，易北河上
The container ship CMA CGM Christophe Colomb on the Elbe at Wedel, Germany

观中发挥了路标的作用。重要的是，环保因素被精心整合进设计方案当中。各种积极的环保措施包括：建筑方位（最大限度的南北阳光入射量）、被动式太阳能策略、大量的保温层和自然通风设计。

海运集装箱的旅程经历了多次转变，而这一趟旅程直接将集装箱引入本文所提到的那些空间创造当中。通过再利用废弃的集装箱，避免了集装箱的废弃，并将它们转变成激动人心的居住空间。尽管从地点、文化和气候方面看，本书所报道的建筑师采用的方法各不相同，但每一个项目都有共同的起点，那就是集装箱。每一份设计纲要与集装箱原先的用途——货物的运送都存在着紧密的联系，从施工阶段到空间安排和场地布置的整个构思过程中都可以看到这一点。

此外，有趣的是，从彼得·库克早期在《建筑电讯》杂志的《插接城市》一文绘制的图纸中，可以看出与这些空间的重新布置和重新利用的现代设计理念之间的联系。这些现代的"插接"空间（海运集装箱）同样很有趣、模块化并且相互连接；在钢制集装箱实用的、横平竖直的背景当中插入创造性的活动和社会政治活动，打造设计巧妙，令人振奋的空间新体验，与集装箱货物运输这一原始用途中的流浪和短暂的特性类似。

should focus on site specificity, considering existing conditions of the current university campus site (including a demolished building site) with a view to the future. The design team infused life into the underused site by using recycled shipping containers. High and low levels are combined with corridors, filled with light and shadow as "spontaneous encounters and play" occur along the route. Views and vistas from the top levels connect not just the users but act as a determined signpost in the wider university landscape. Importantly, environmental factors have been carefully integrated into the scheme. Location (maximum north-south exposure), passive solar strategies, abundant insulation and natural ventilation add to the positive environmental trail.
The journey of the shipping container has taken many turns, and this journey has lead directly to the creation of spaces such as those covered in this article. By reusing the dumped containers, the containers have been saved from disuse and been transformed into exciting, habitable spaces. Although the architects featured here have all taken very different approaches, considering the variation of location, culture and climate, each project has the starting point in common: the shipping container. Each brief has a strong link to the original use, delivery of cargo, and this can be seen in the construction stages, and throughout the ideation in terms of spatial arrangements and placement on site.
Furthermore, interestingly, references to the early drawings of Peter Cook for *Archigram*'s *Plug-in City* can be likened to these contemporary ideas of rearrangement and reuse of space. These present-day "plug-in" spaces (shipping containers) are similarly playful, modular, interlinking; plugging into the creative and socio-political activities amidst the practical, straight edged backdrop of the steel container to well-designed, exciting new spatial experiences, akin to the nomadic and ephemeral of the original use: the transportation of goods. Heidi Saarinen

154

海运集装箱的改造再利用 Re-use / Re-model – the Shipping Container

城市船舱
BIG

最近几年，整个丹麦的学生申请人数量持续大规模地增加。随着学生数量的持续增加，需要新增更多的学生公寓。允许城市扩张的对策少之又少。而哥本哈根的海港一直是城市中心未被充分利用和未被开发的区域。通过引进一种经过优化的、最适合海港城市的建筑类型，我们可以推出一种住房解决方案，使学生可以留在城市的中心。同时，标准化的集装箱系统使得货物可以通过陆路、水路或航空等途径，利用复杂的运营商网络以较低的成本运往世界各地。通过利用标准的集装箱系统，我们可以为极其灵活的建筑类型提供框架。将9个集装箱堆叠成一个圈，就形成了绕着中央的冬日花园的15个工作室型公寓；冬日花园是学生们日常碰面的地方。这种公寓还是漂浮的，像船只一样，所以在其他需要经济实惠的公寓但空间不足的海港城市，也能复制建造出来。

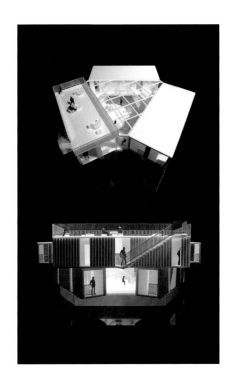

CREATING THE COURTYARD
We then tried arranging the containers into a triangular composition to frame a central courtyard. This allowed us to minimize the footprint of the pontoon, while opening views to the water – optimizing the housing unit.

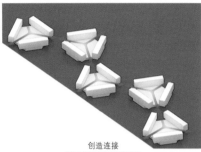

CREATING CONNECTIONS
By detaching the corners slightly, a hexagonal courtyard with open corners takes form – creating a connection between communities and allowing for further expansion.

MULTI-LEVEL CONNECTIONS
Another layer of container units completes the circle, forming a hexagon of overlapping entities.

INTERNAL COMMUNITIES
Courtyards at the heart of the Urban Rigger create opportunities for community activity within each unit. As weather in Denmark changes drastically from season to season, we enclose the gaps with greenhouse glass, minimizing the thermal exposure during the winter months to enclose the largest possible amount of space with the minimal amount of surface.

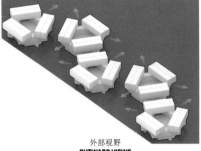

OUTWARD VIEWS
The hexagons combine in a variety of ways to form different communities, focusing views in all directions and creating clear and unique views directed toward the sea.

Urban Rigger

Recent years have demonstrated a substantial and sustained increase in the number of student applicants throughout Denmark. As the number of students continues to grow, additional student housing will be needed to accommodate them. There are few strategies that allow cities to expand. Yet, Copenhagen's harbor remains an underutilized and underdeveloped area at the heart of the city. By introducing a building typology optimized for harbor cities we can introduce a housing solution that will keep students at the heart of the city. Meanwhile, the standardized container system has been developed to allow goods to be transported by road, water or air, to anywhere in the world in a complex network of operators at a very low cost. By making use of the standard container system we can offer the framework for an extremely flexible building typology. By stacking 9 container units in a circle, we can create 15 studio residences which frame a centralized winter garden; this garden is used as a common meeting place for students. The housing is also buoyant, like a boat, so that it can be replicated in other harbor cities where affordable housing is needed, but space is limited.

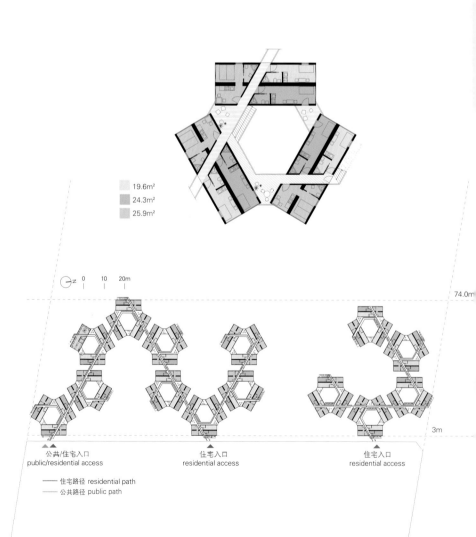

项目名称：Urban Rigger
地点：Copenhagen, DK
建筑师：BIG
主管合伙人：Bjarke Ingels, Jakob Sand
项目总监：Joos Jerne
项目团队：Aaron Hales, Adam Busko, Agne Tamasauskaite, Aleksandra Sliwinska, Andreas Klok Pedersen, Annette Birthe Jensen, Birgitte Villadsen, Brage Mæhle Hult, Brigitta Gulyás, Carlos Soria, Christian Bom, David Zahle, Dimitrie Grigorescu, Edda Steingrimsdottir, Edmond Lakatos, Elina Skujina, Finn Nørkjær, Ioana Fartadi Scurtu, Jacob Lykkefold Aaen, Jakob Lange, Kamila Rawicka, Lise Jessen, Lorenzo Boddi, Magdalene Maria Mroz, Nicolas Millot, Perle van de Wyngert, Raphael Ciriani, Stefan Plugaru, Stefan Wolf, Tobias Hjortdal, Toni Mateu, Tore Banke, Viktoria Millentrup
合作方：BIG Ideas, Danfoss A/S, Grundfos DK A/S, Hanwha Q CELLS Ltd., Miele, NIRAS A/S, Dirk Marine/House on Water
客户：Udvikling Danmark A/S
用途：Housing / 建筑面积：680m²
竣工时间：2016.9.20
摄影师：courtesy of the architect
(except as noted)

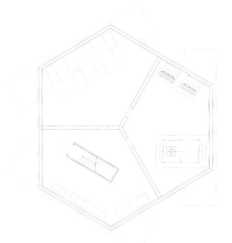
一层 ground floor

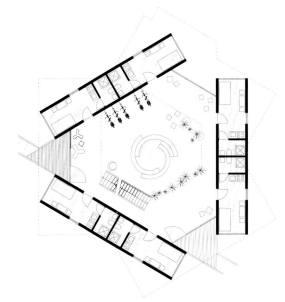
二层 first floor

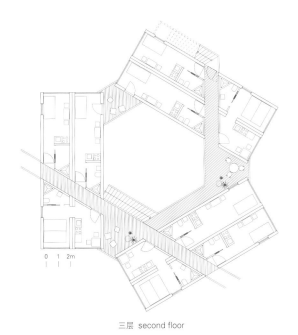
三层 second floor

屋顶 roof

Ccasa青年旅店
TAK Architects

Ccasa青年旅店位于越南芽庄，是首家由集装箱改造而成的青年旅店。青年旅店的位置在城市北部，距离市中心约3km，步行去海滩只要3分钟。它邻近芽庄的著名景点——Hon Chong – Hon Vo（与迷人的当地神话有关的巨大岩岬）、高棉女神Ponaga庙以及天然温泉度假胜地I-resort。

这是一家为背包客而建造的青年旅店，"四海一家"是旅店的设计箴言。Ccasa青年旅店在功能上像是一个家庭住宅，集装箱内有双层床作为卧室，还有公共厨房和客厅，露台屋顶有娱乐室，还有洗涤区域、卫生间和浴室。因此，床位空间被减到最小，仅够睡觉用。相比之下，共享空间被扩大到最大，以增加住客之间的交流。洗涤区域同样为共享空间。

旅店由三个功能区组成：服务区、休息区和盥洗区。服务区由钢框架和黑色金属板构成。休息区设置在三个旧的海运集装箱中，集装箱被涂成三种颜色，代表三种类型的卧室。盥洗区以标准形式建造，采用白色涂料的毛面砖和混凝土。这三个区域通过公共空间（同时也是共享空间和交通空间）相连，这个公共空间最大程度地对外开放，引入新鲜空气和自然光。正因如此本项目才变得那么舒适和谐。

该项目的其他特点包括通向卧室的入口，不再是闷热的走廊，而是开放明亮的连桥，绿树和藤架掩映其间。所以当住客进出卧室的时候，会感到非常轻松和舒适。除此之外，露台屋顶也采取了大胆的手法，悬挂于上空空间的大吊床给人一种漂浮于空中的感觉。

通过使用老旧的集装箱、钢架、绿树以及藤架，Ccasa青年旅店体现出一种强烈的工业风，但又不失和谐自然的氛围。此外，水泥彩瓦、老木窗、扁平的簸箕、粗水泥被巧妙地用来吸引人们的注意力，让人不禁回忆起一些越南传统建筑的元素。旅店周围的绿色藤架不仅使建筑看起来更富有绿意，还成为建筑的第二层表皮，防止阳光直射，给室内降温。

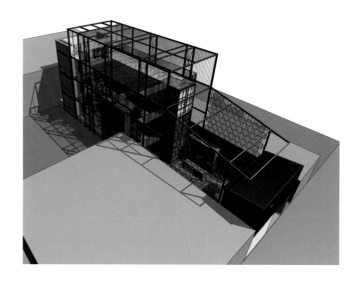

项目竣工后，Ccasa青年旅店已然成为深受旅客欢迎的旅游目的地。它还为城市增加了更多的绿色空间，有助于保护环境。

Ccasa Hostel

Ccasa Hostel in Nha Trang, Vietnam, is the first hostel to be built from shipping containers. The hostel is located on the north of the city, about 3km from the center, a three-minute walk from the beach. It is close to famous sites in Nha Trang: Hon Chong – Hon Vo (huge rocky headland associated with a fascinating local myth), Khmer's temple of the goddess Ponaga, and I-resort, a natural mineral water spring.

This is a hostel for backpackers, created on the motto that everyone in the world is in one big family. Ccasa is built to function like a family home, with cabin beds inside containers as bedrooms, shared kitchen and living room, a playroom on the terraced roof, washing area, toilets and bathrooms. Therefore, the bed space was reduced to the minimum, just enough to sleep. In contrast, the shared space was expanded to the maximum to increase the connection among travelers. The washing area is also shared.

The hostel consists of three functional blocks: one for serving, one for sleeping and one for washing. The serving block was made of steel frames and black painted metal sheets. The sleeping block was set in three old shipping containers

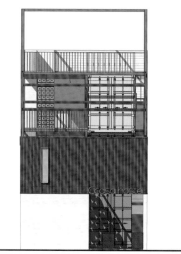

东立面 east elevation

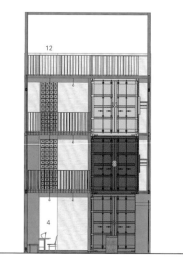

A-A' 剖面图 section A-A'

1.接待处	5.六人间宿舍	9.走廊	1. reception	5. cabin-6 beds dorm	9. corridor
2.停车场	6.卫生间+淋浴间	10.储藏室	2. parking area	6. toilet + shower	10. store
3.开放厨房	7.洗衣房	11.家庭房	3. open kitchen	7. laundry	11. family cabin
4.公共空间	8.四人间宿舍	12.空中吊床	4. common space	8. cabin-4 beds dorm	12. sky hammock

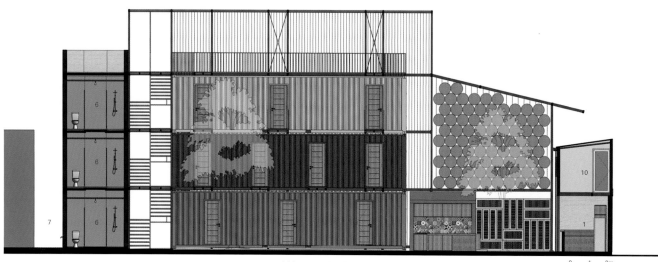

B-B' 剖面图 section B-B'

that was painted three colors to symbolize three types of bedrooms. The washing block was built in a normal way with white painted rustic bricks and concrete. These three blocks were connected by the common space that is shared space as well as traffic space that is opened to the maximum to let in fresh air and sunlight. This is the main feature that makes the project so soft and harmonious.

Some other features of this project are the entrances to the bedrooms which are no longer stuffy corridors, but instead are open luminous bridges covered by green trees and pergola, so that travelers will feel very relaxed and comfortable when entering or leaving the bedroom. The terraced roof also takes a bold approach with the large hammocks hung across the void to create the feeling of floating in the air.

By using old shipping containers, steel frames, green trees, and the pergola, Ccasa Hostel has evoked a strong, industrial yet peaceful natural feeling. In addition, encaustic cement tiles, old wood windows, flat winnowing baskets, rustic cement were used cleverly to attract attention and recall elements of traditional Vietnamese architecture. The pergola covering the hostel not only makes it look greener but also acts as a second skin to protect it from direct sunlight and cool the air inside.

Since its completion, Ccasa Hostel has become a popular destination for the travelers. It also adds green space to the city and helps to protect the environment.

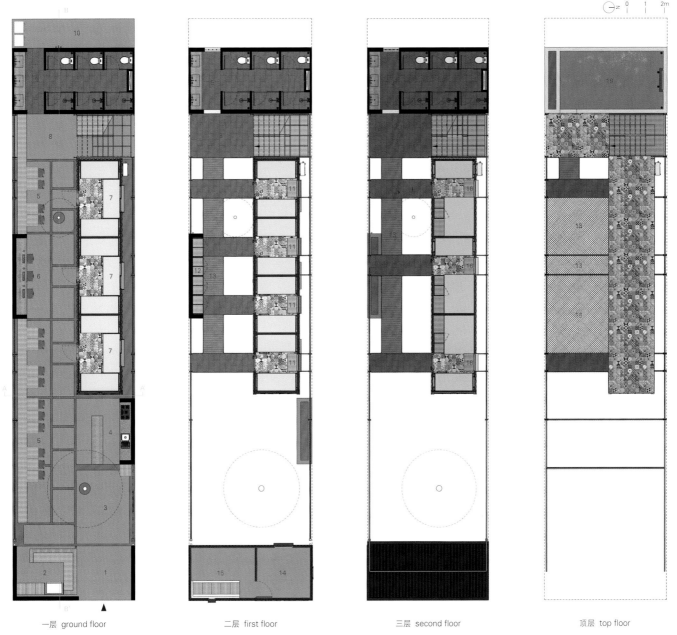

一层 ground floor　　二层 first floor　　三层 second floor　　顶层 top floor

1.接待门厅 2.接待处 3.停车场 4.开放厨房 5.公共空间 6.商务中心 7.六人间宿舍 8.门厅 9.卫生间+淋浴间 10.洗衣房 11.四人间宿舍 12.储物柜 13.走廊 14.员工房 15.储藏室 16.家庭房 17.多功能休闲区 18.空中吊床 19.技术设备区

1. reception lobby 2. reception 3. parking area 4. open kitchen 5. common space 6. business center 7. cabin-6 beds dorm 8. lobby 9. toilet+shower 10. laundry 11. cabin-4 beds dorm 12. locker 13. corridor 14. staff room 15. store 16. family cabin 17. multipurpose relaxing area 18. sky hammock 19. technical area

项目名称：Ccasa Hostel / 地点：40 Sao Biên, Nha Trang, Khanh Hoa, Vietnam / 建筑师：TAK Architects / 总建筑师：Ngo Tuan Anh
总建筑面积：195m² / 结构：steel, concrete / 材料：steel, concrete, brick, wood / 竣工时间：2016 / 摄影师：©Quang Tran (courtesy of the architect)

艾杰大学的高新科技园
ATÖLYE Labs

海运集装箱的改造再利用 Re-use / Re-model – the Shipping Container

由于新的生产关系在教育、研究和工业之间形成，iDE EGE高新科技园A.Ş.邀请ATÖLYE Labs在土耳其西部新兴的大都市伊兹密尔中心的科技园设计了一座用于促进跨学科交流合作的公共设施建筑。该项目最终由35个二手集装箱构成，该中心将成为伊兹密尔的艾杰大学乃至整个爱琴海地区崭新而充满活力的研究型社区。

占地1000m²的科技园将为那些致力于生物技术、能源、材料和软件研究的土耳其大公司以及国际公司提供独立的研究和开发场所。

该科技园建筑的独特之处不仅在于其快节奏的研究、设计和施工过程（所有的项目都在短短的九个月内完成），还因为项目的设计纲要、场地和规划都是由设计团队自己开发和改进的。此外，因地制宜的、环保的、适应未来发展的主要设计原则也有助于为土耳其乃至全世界其他类似机构的设计创建一个设计样板。

在偌大的大学校园内发现了一处堆放拆除建筑残垣断瓦的闲置场地，本项目以此为起点开始创建。通过对来自12km之外伊兹密尔港的二手集装箱的再利用，设计团队可以通过未经充分利用的建筑材料来升级改造一个未经充分利用的场地。

场地调查、建筑朝向、现有的校园交通流线、风向、树荫区域和先前建筑的轮廓，这些都有助于建立有意义的、经济可行的规划分区、体量安排，并最终形成一个流畅的使用者交通流线。

除了循环和再利用材料的大量使用，该项目还展现了广泛的生态策略。在集装箱的摆放上，使南北向的立面尽最大化接触阳光，设计最窄小的横剖面，这样可以使被动式太阳能和自然通风两个生态策略发挥到最大化。

现存的树木、优化设计的遮阳设施、南向的阳光控制镀膜玻璃窗、厚实的保温层、高效的空气调节装置、软木橡树皮等自然材料以及LED照明系统都有利于将建筑对自然环境的影响降至最低。

从长远来看，该建筑的适应性和灵活设计是其最大的资产，该项目突出了多项技术细节，比如，裸露的梁柱、可见的电子托盘、充足的插座、高容量的通风设备、局部可控的供暖和制冷系统，以及起到支持作用的子结构，都有助于实现未来的分隔系统。随着时间的流逝，所有这些系统都有利于轻而易举地修改空间计划。该项目成为该类型建筑的一个示范，在伊兹密尔乃至整个世界范围内，它势必会刺激形成一个人才汇集的社区。

建筑师：ATÖLYE Labs / 设计团队：Engin Ayaz, Nesile Yalçın, Nujen Acar, Buşra Tunç / 项目经理：Nesile Yalçın / 项目团队：ATÖLYE Labs_Strategic consulting, Sustainability consulting, Architectural design, Project management, Site control; Antre Design_Architectural consulting; Venta Mühendislik_Mechanical; Sinapsen Elektrik_Electrical; Methal Mühendislik_Static analysis; Parça Proje_Lighting consulting; Murathan Sırakaya, Gökhan Gürbüz_Rendering / 客户：IDE EGE Teknopark A.Ş. / 摄影师：©Yerçekim Photography (courtesy of the architect)

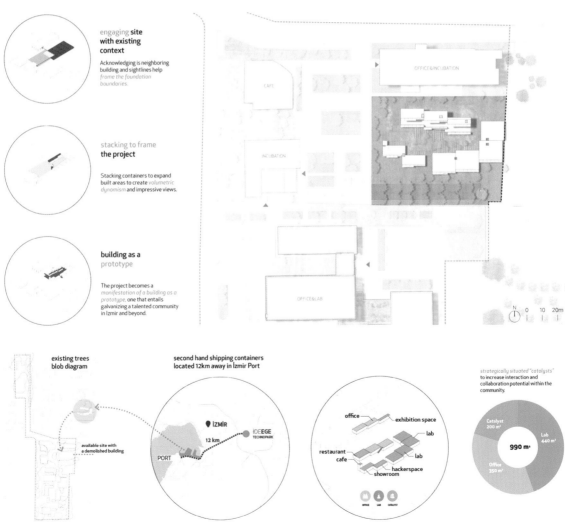

项目名称: Ege University Technopark / 地点: Ege Üniversitesi Teknopark, Turkey / 用途: offices, laboratory space, gallery, cafe, eatery restaurant, showroom, gallery, exterior terrace and garden / 范围: Core & shell office building (architecture + landscape architecture + engineering + spatial graphics + branding) designed with 35 second-hand shipping containers / 总建筑面积: 1,000m² / 园林景观面积: 800m² / 竣工时间: 2015.10

Ege University Technopark

New productive relationships between education, research, and industry, iDE EGE Technopark A. Ş. approached ATÖLYE Labs about a facility to foster interdisciplinary collaboration and anchor the emerging technopark in the heart of Izmir, a bustling cosmopolis in Western Turkey. The project repurposes 35 second hand shipping containers to form the centerpiece of a new, vibrant research community on the campus of Ege University, Izmir and Aegean region at large.
The 1000m² technopark facility will house independent R & D facilities of large Turkish and International companies focusing on biotechnology, energy, materials and software research. The technopark facility was unique not only in terms of its fast-paced research, design and construction process (all in all in a tight 9-month schedule), but also because its brief, site and program were developed and refined by the design team itself. Furthermore, key design principles of site-specificity, ecology and future proofing helped create a role model for similar institutions in Turkey and beyond.

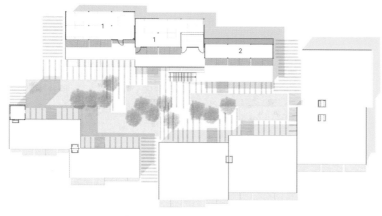

1. 办公室 2. 展览空间 1. office 2. exhibition space
二层 first floor

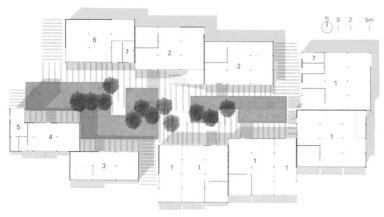

1. 实验室 2. 办公室 3. 创客空间 4. 展室 5. 咖啡厅 6. 餐厅 7. 卫生间
1. lab 2. office 3. hacker space 4. showroom 5. cafe 6. restaurant 7. wc
一层 ground floor

The project started with the discovery of a dormant site with the rubble of a demolished building amidst the large university campus. By repurposing locally acquired second hand shipping containers located 12km away in Izmir port, the design team was able to upcycle an underused site with underused construction materials.

Looking into the site, solar orientation, existing campus circulation routes, wind angles, tree shaded areas and the contours of the previous building helped craft a meaningful and financially viable programmatic division, volumetric arrangement and ultimately a fluid user circulation.

Aside from an exceptional amount of material recycling and reuse, the project exhibits a wide gamut of ecological strategies. By placing container modules with maximum north-south exposure and narrow cross sections, the design maximized the ability to use passive solar strategies coupled with natural ventilation.

Existing trees, optimally designed shading devices, solar coated southern windows, thick insulation, efficient air conditioning, natural materials such as cork, and LED lighting systems all helped minimize the building's environmental impact.

Given that adaptability and resilience of a core & shell building are its biggest assets in the long term, the project features multiple technical details such as exposed beams and columns, visible electrical trays, abundant plugs, high-capacity ventilation, locally controllable heating-cooling systems and supporting sub-structure to help build separators in the future. All these systems help easily modify spatial programs over time. The project becomes a manifestation of a building as a prototype, one that entails galvanizing a talented community in Izmir and beyond.

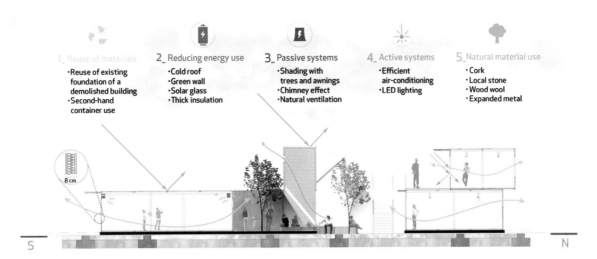

生态设计原理
ecological design principles

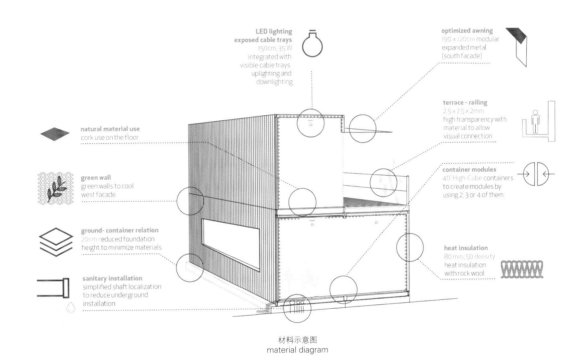

材料示意图
material diagram

1956年，一名美国公路运输经营者发明了集装箱。如今，世界上80%的货物（按价值）通过集装箱运送，每年集装箱的制造量大约为160万。2010年，在世界范围内处于使用中的运输集装箱数量超过了1800万。

集装箱为这家当代音乐中心的布局设计方面提供了极大的灵活性，演出空间可以根据在那里举办的活动和所需的不同类型的氛围进行组织。可以增加额外的房间，以补充符合设计规范要求的排练室。

这座新建的当代音乐中心位于锡安南部一处先前为工业厂房的场地内。音乐大厅被命名为"自由港"，其位置特别适合举行夜场活动；所选的工业区完美解决了噪声问题。本建筑包括两个各不相同却又互为补充的设计：一个可以容纳400人的音乐大厅和为当地乐队设计的排练空间。对二手海运集装箱的使用实现了一种经济的设计方式。集装箱系统的模块化在组织方面提供了极大的灵活性并且实现了快速完工。集装箱材料的强度能够很好地保护建筑不受到破坏，并经得起房屋的集约化使用。最后，这些带有醒目视觉标志的模块的回收利用为这种新式的文化功能营造了一种强大而又始终如一的形象。

Sion Music Center

In 1956, an American road haulage operator invented the container. Today, containers transport 80% of goods (by value), and around 1.6 million of them are constructed every year. In 2010, there were over 18 million containers in use worldwide for transport.

A container gives great flexibility in the arrangement of the contemporary music center. The performance space can be organized according to the activities taking place there and

锡安音乐中心

Savioz Fabrizzi Architectes

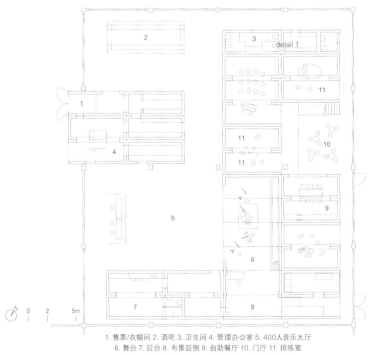

1. 售票/衣帽间 2. 酒吧 3. 卫生间 4. 管理办公室 5. 400人音乐大厅
6. 舞台 7. 后台 8. 布景后侧 9. 自助餐厅 10. 门厅 11. 排练室
1. ticketing, cloakroom 2. bar 3. toilet 4. administration 5. 400 people concert hall
6. stage 7. backstage 8. behind the scenes 9. cafeteria 10. foyer 11. rehearsal room
一层 ground floor

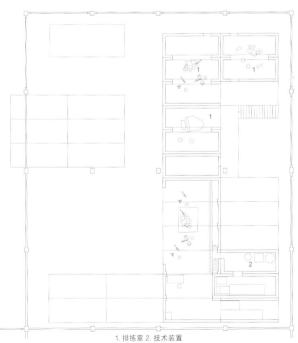

1. 排练室 2. 技术装置
1. rehearsal room 2. technical installation
二层 first floor

the different types of ambiance required. Extra rooms can be added, supplementing the rehearsal rooms required by the project specifications.

The new contemporary music center is located in a former industrial hall in the south of Sion. The concert hall called "le port-franc" enjoys a particularly favourable situation to its nocturnal activities; the industrial area chosen is ideal for the noise problem. The building contains two different and complementary programs: a 400 people concert hall and rehearsal spaces for local bands. The use of second-hand shipping containers allows an economy of means. The modularity of the system provides great flexibility in the organization and enables a rapid execution. The resistance of container materials offers an ideal protection against vandalism and intensive use of the premises. Finally, the recycling of these modules with marked visual identity creates a strong and consistent image for this new and alternative pole of culture.

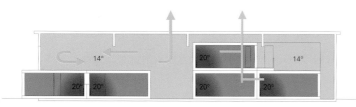

热能利用示意图 thermical diagram

声效示意图 acoustic diagram

项目名称：Sion Music Center
地点：Sion, Switzerland
建筑师：Savioz Fabrizzi Architectes
电子工程师：Domotech Systems
土木工程师：Alpatec sa, Martigny
暖通工程师：Technitherm, Yannick Rossier
声效工程师：BS Bruno Schroeter
功能：concert hall, repetition rooms, loge, ticket desk, bar
总建筑面积：1,100m²
竣工时间：2011.12
施工时间：2014—2015
摄影师：©Thierry Sermier
(courtesy of the architect)

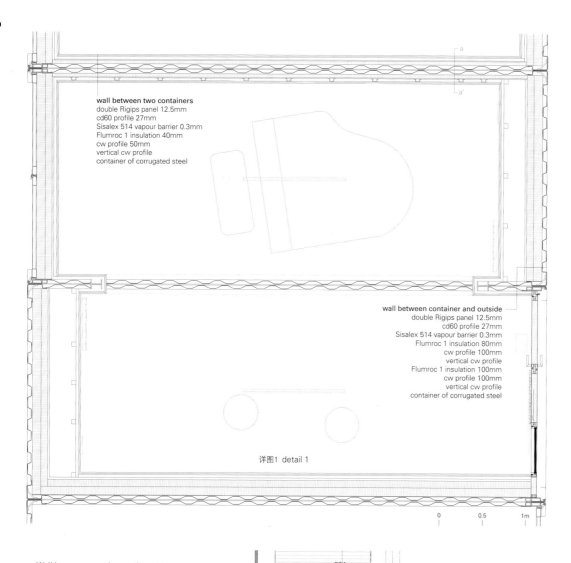

wall between two containers
double Rigips panel 12.5mm
cd60 profile 27mm
Sisalex 514 vapour barrier 0.3mm
Flumroc 1 insulation 40mm
cw profile 50mm
vertical cw profile
container of corrugated steel

wall between container and outside
double Rigips panel 12.5mm
cd60 profile 27mm
Sisalex 514 vapour barrier 0.3mm
Flumroc 1 insulation 80mm
cw profile 100mm
vertical cw profile
Flumroc 1 insulation 100mm
cw profile 100mm
vertical cw profile
container of corrugated steel

详图1 detail 1

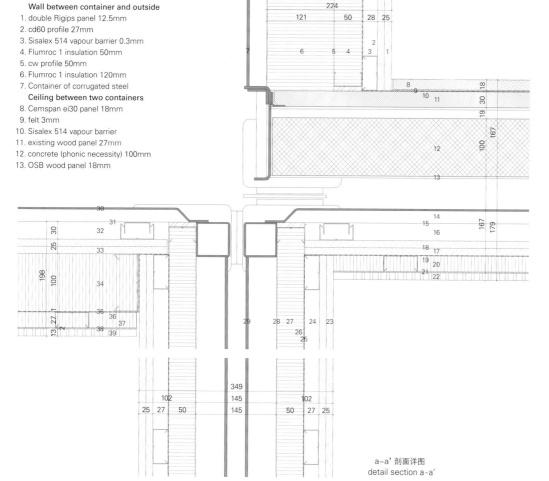

Wall between container and outside
1. double Rigips panel 12.5mm
2. cd60 profile 27mm
3. Sisalex 514 vapour barrier 0.3mm
4. Flumroc 1 insulation 50mm
5. cw profile 50mm
6. Flumroc 1 insulation 120mm
7. Container of corrugated steel

Ceiling between two containers
8. Cemspan ei30 panel 18mm
9. felt 3mm
10. Sisalex 514 vapour barrier
11. existing wood panel 27mm
12. concrete (phonic necessity) 100mm
13. OSB wood panel 18mm

Ceiling between two containers
14. container of corrugated steel
15. felt 3mm
16. cd30 profile 30mm
17. double ei30 Rigips panel 12.5mm
18. Sisalex 514 vapour barrier 3mm
19. Flumroc 3 phonic insulation 27mm
20. cd profile 27mm
21. fibreglass
22. black perforated Rigips panel 12.5mm

Wall between two containers
23. double Rigips panel 12.5mm
24. cd60 profile 27mm
25. Sisalex 514 vapour barrier 0.3mm
26. Flumroc 1 insulation 40mm
27. cw profile 50mm
28. vertical cw profile
29. container of corrugated steel

Ceiling between container and outside
30. container corrugated steel
31. felt 3mm
32. cd30 profile 30mm
33. double ei30 Rigips panel 12.5mm
34. Flumroc 1 insulation 100mm
35. Sisalex 514 vapour barrier 3mm
36. Flumroc 3 phonic insulation 27mm
37. cd profile 27mm
38. fibreglass
39. black perforated Rigips panel 12.5mm

a-a' 剖面详图
detail section a-a'

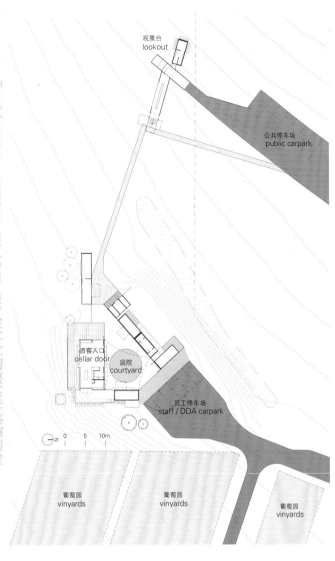

魔鬼之角

Cumulus Studio

魔鬼之角酒窖坐落于澳大利亚塔斯马尼亚东海岸的观光游览道路旁，新建成的酒窖入口观景台位于当地最大的一家葡萄酒庄园里，在这里可将弗雷西内半岛的景色尽收眼底。2015 年 12 月重新开放的这个为布朗兄弟酒庄而设计的项目试图扩大对此处有代表性的景观的体验，来创造全新的塔斯马尼亚东海岸的旅游体验。酒窖入口原先只是一座小型的可拆卸建筑，经过扩建，增加了观景台和免费食物体验区，并为季节性活动提供了空间。酒窖入口观景台被设计成一座外观由木板条覆盖的建筑，通过相似的美观度和材料处理，形成了对历史悠久的传统农场／乡村居所的现代诠释。酒窖入口和食品市场环绕着庭院空间布置，庭院提供了面向周边环境的遮蔽和休息区，同时允许视线穿过品尝区眺望到远处的 Hazards 山脉以及通往开放平台区的通道。对一系列木材覆面集装箱的精心排布形成了特殊的观景图框，使参观者可以欣赏到葡萄园内部和外部的美丽景色。观景台是设计中的关键部分，不仅为居住区提供了一个视觉上的标志物，同时也是对魔鬼之角葡萄酒原产地景观的一种诠释。对酒的品鉴可以通过各种各样细微的口感来判断，同样，此处的景观也可以通过许多不同的方式来欣赏。观景台的设计正是考虑到了以上理念。三个与众不同的空间代表了建筑场地上的三种不同而又独特的景观：首先是"天空"，接着是"地平线"，而最后是一座蜿蜒向上的"塔"，从塔的各个方位都能欣赏到景色，最后到达塔顶，广阔的海湾美景一览无余。通过打造富有活力的观景台并提供相关服务设施，本项目成功地吸引了参观者进入这个新升级改造的魔鬼之角葡萄酒酒窖。

Devil's Corner

Located on the scenic drive along Tasmania's East Coast, the new Devil's Corner Cellar Door and Lookout sits within one of Tasmania's largest vineyards, with a panoramic view over the Freycinet Peninsula. Reopened in December 2015, this project for Brown Brothers seeks to amplify the experience of this iconic view to create a new tourism experience on the East Coast of Tasmania. Originally a small demountable building, the Cellar Door has been extended and expanded, paired with a lookout and complimentary food experiences, providing a backdrop for seasonal events. The Cellar Door

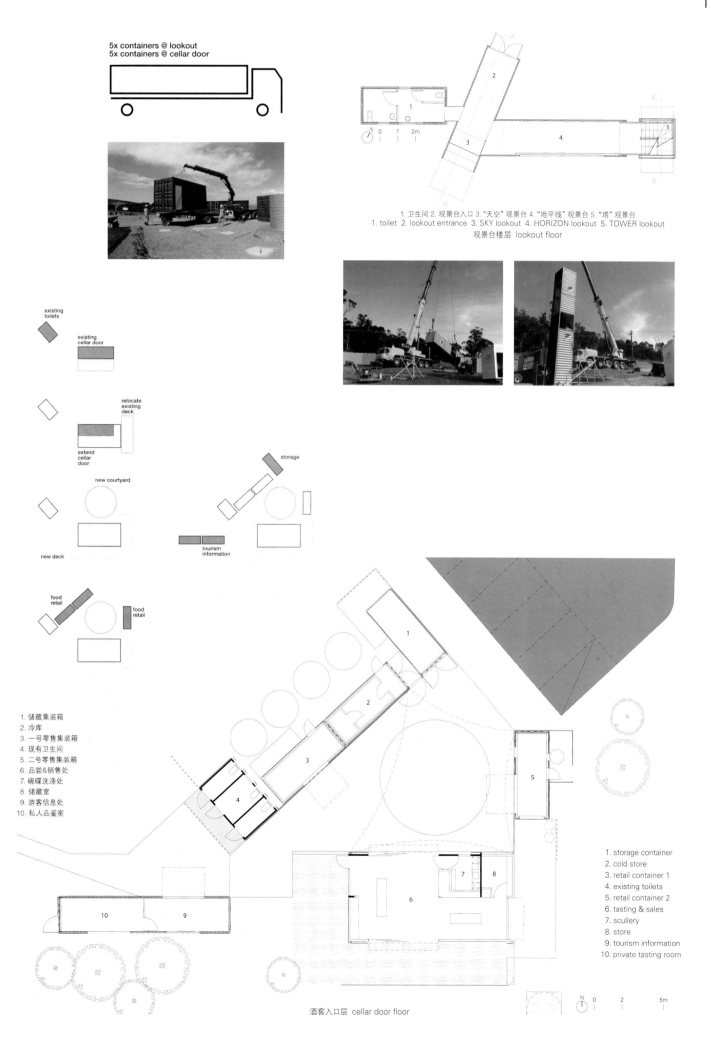

and Lookout were designed as a loose collection of timber clad buildings that, through similar aesthetic and material treatment, form a modern interpretation of traditional farm / rural settlement that gather over time. The Cellar Door & food market have been collected around a courtyard space which allows shelter and respite from the surrounding environment, while allowing views through the tasting space to the Hazards beyond and access to open deck spaces. Through the careful placement of a series of timber clad shipping containers, visitors are invited to visually explore the landscape within and around the vineyard through curated framed views. The lookout element is a critical component of the design, not only in providing a visual signifier for the settlement but also as a way of interpreting the landscape from which the Devil's Corner wines originate. In the same way that an appreciation of wine can be gained through understanding its subtleties and varying "in-mouth" sensations, there are many ways landscape can be appreciated. The lookout plays with this idea. The three distinct spaces reference different and unique views of the site – firstly the SKY, then the HORIZON and lastly the TOWER which winds its way upward providing views to each of the compass points before culminating in an elevated and expansive view of the bay. By creating a dynamic scenic lookout and providing associated facilities, visitors are drawn to the new upgraded cellar door for the Devil's Corner wine label.

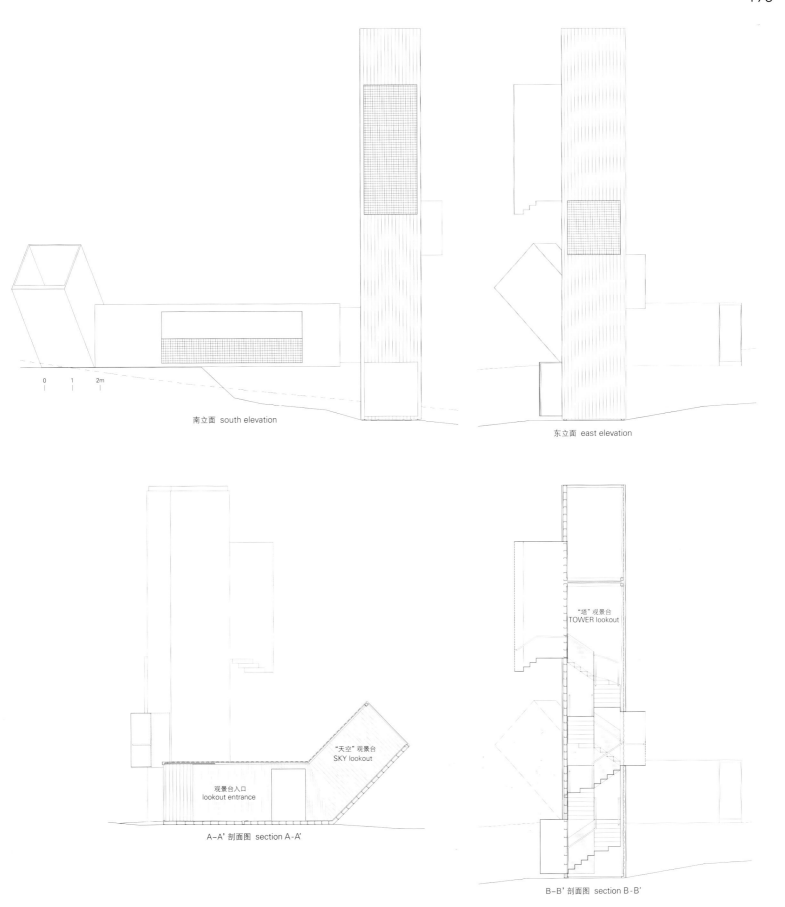

南立面 south elevation

东立面 east elevation

A-A' 剖面图 section A-A'

"天空"观景台 SKY lookout

观景台入口 lookout entrance

"塔"观景台 TOWER lookout

B-B' 剖面图 section B-B'

项目名称：Devil's Corner / 地点：Sherbourne Road, Apslawn, Tasmania Australia 7190
建筑师：Cumulus Studio _ Peter Walker, Liz Walsh, Andrew Geeves, Fiona McMullen, Todd Henderson
结构工程师：Aldanmark / 建筑测量师：Castellan Consulting / 旅游顾问：Simon Currant
酒店顾问：David Quon / 环境：Red Sustainability Consultants / 建造商：Anstie Constructions
建筑面积：572m² (Lookout: 112m², Cellar door & Market area: 460m²) / 施工时间：2015.8 — 2015.12 / 竣工时间：2015
摄影师：©Tanja Milbourne (courtesy of the architect) (excepted as noted)

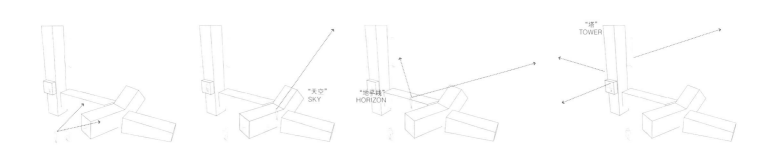

"天空" SKY "地平线" HORIZON "塔" TOWER

P30
Sou Fujimoto Architects
Sou Fujimoto graduated from the Department of Architecture, Faculty of Engineering at the Tokyo University. He established "Sou Fujimoto Architects" in 2000. His recent work, Mille Arbres won the competition of Reinventer Paris with Manal Rachadi OXO in 2016. He also won the 1st prize in the New Learning Center at Paris-Saclay's Ecole Polytechnique in 2015. His representative works are "the Serpentine Gallery Pavilion 2013" (2013), "House NA" (2011), "Musashino Art University Museum & Library" (2010) and so on.

P182
Savioz Fabrizzi Architectes
Founded in 2004 by the two architects, Laurent Savioz and Claude Fabrizzi, is trying to respond with the best conditions to the needs of the clients by providing all the architectural services from the project to the achievement. Their work is based on the analysis of a site in its natural or built state in order to identify the essentials elements that could enhance, preserve or qualify a site. In this way, the firm enhances the cultural role of the architecture based on the analysis of a function, respectively a program, its place in the history and the culture of a region.

P126
Schmidt Hammer Lassen Architects
Was founded in Aarhus, Denmark in 1986 by architects Morten Schmidt, Bjarne Hammer and John F. Lassen. Is one of Scandinavia's most recognized and award-winning architectural practices with 30 years of experience. Now has Senior Partners Kim Holst Jensen and Kristian Lars Ahlmark, Partners Chris Hardie and Rong Lu. Working out of studios located in Copenhagen, Aarhus, Shanghai and London, they are deeply commited to the Nordic architectural traditions based on democracy, welfare, aesthetics, light, sustainability and social responsibility. Received prestigious, national and international awards through the years including the Governor General's Medal in Architecture in Canada/A+Award 2016/Årets Bygge 2016/Architectural Review MIPIM Future Project Awards 2015/World Green Design Product Award 2014/RIBA National Award 2013/RIAS Award 2013/ArchDaily Building of the Year Award 2011/LEAF Award 2011.
Rong Lu, Chris Hardie, Kim Holst Jensen, John F. Lassen, Kristian Lars Ahlmark, Morten Schmidt, Bente Damgaard, Bjarne Hammer, from left.

Cumulus Studio

Co-founded by Todd Henderson[left] and Peter Walker[right], the studio is an award-winning architecture practice with offices in Hobart(led by Peter), Launceston(led by Todd) and Melbourne. The three offices operate as one combined studio, consist of 15 designers, including 9 registered architects, and specialize in tourism, residential, commercial, heritage, urban design and interior architecture. From the Latin word for "heap" or "pile", Cumulus believes that through working together a critical mass of ideas can be accumulated quickly - forming idea clouds. These ideas can take on any size or form and be manipulated to suit conditions. Cumulus does not have a stylistic approach or predefined ways of working, rather preferring that each new architectural design emerges from the unique conditions of the client and site.

ATÖLYE Labs

Engin Ayaz[left] is the cofounder of ATÖLYE Labs and a multidisciplinary designer, experienced in Architecture, Engineering and Interaction Design. He studied Civil and Environmental Engineering and Architecture Design at the Stanford University. He worked in several offices in Europe and US as a sustainable design consultant. He received a master's degree of Interaction Design and Media Arts at NYU Tisch School of the Arts in 2013. Nesile Yalçın[right] is an architect based in Istanbul and specialized in designing new generation office, learning, research spaces and campuses. She graduated from the ITU school of architecture and got her M. Arch at the ITU by studying on Investigating environmental behavior of architects in the context of sustainability.

©Jessica Van Campen

P114
Carquero Arquitectura
Carlos Quevedo Rojas graduated from the E.T.S. Arquitectura of Seville and received his Master's Degree in Architecture and Historical Heritage. He received another Master's Degree from the University of Granada and studied "Art History: Investigation and Guardianship of the Heritage". His research works were awarded like the Urban Planing National Prize "Ricardo Santos Díez" and so on. His works and projects try to delve the conceptual principles of the architecture and the intervention in the heritage. He is currently a professor at the School of Art and Design of Cádiz.

P162
TAK Architects
Was founded by Ngo Tuan Anh. He was born in Thai Binh province in 1984. He graduated from the Ho Chi Minh City University of Architecture. He worked in Transform Architecture for 2 years and spent 5 years as a project manager for several large projects.

P110, P138
OMA
Is an international practice operating within the traditional boundaries of architecture and urbanism. AMO, a research and design studio, applies architectural thinking to domains beyond. Partner of OMA, Reinier de Graaf, joined OMA in 1996. He is responsible for building and masterplanning projects in Europe, Russia, and the Middle East including Holland Green in London(2016), the new Timmerhuis in Rotterdam(2015) and De Rotterdam(2013). In 2002, he became the director of AMO, the think tank of OMA. He recently curated two exhibitions, On Hold at the British School in Rome in 2011 and Public Works: Architecture by Civil Servants (Venice Biennale, 2012; Berlin, 2013).

©Ekaterina Izmestieva

P154
BIG
Was founded in 2005 by Bjarke Ingels and now based in Copenhagen and New York. The group of architects, designers, builders, and thinkers are operating within the fields of architecture, urbanism, interior design, landscape design, product design, research and development. The office is currently involved in a large number of projects throughout Europe, North America, Asia and the Middle East. Their architecture emerges out of a careful analysis of how contemporary life constantly evolves and changes.
Kai-Uwe Bergmann, Andreas Klok Pedersen, Brian Yang, Daniel Sundlin, David Zahle, Beat Schenk, Bjarke Ingels, Sheela Maini Søgaard, Jakob Lange, Thomas Christoffersen, Finn Nørkjær, Jakob Sand, from left, in the picture p.200 bottom.

Allies and Morrison
Allies and Morrison's work ranges from architecture, interior design and conservation to urban planning, consultation and research. Partner of Allies and Morrison, Simon Fraser, studied at the Cardiff School of Art and subsequently worked on transport and retail interiors in London, Spain and Finland before joining Allies and Morrison in 1998. He was appointed partner in 2013. He has extensive experience of working on large and complex projects, often in historically sensitive settings and he is currently the Project Partner for a high specification residential development, delivering 58 apartments across three new blocks on the edge of Holland Park in Kensington, London which involves the refurbishment of the Grade II* listed, former Commonwealth Institute for the Design Museum.

P10
The Norman Foster Foundation
Is based in Madrid, Spain. Promote the importance of architecture, infrastructure and urbanism to serve society through experimental and research projects. Some of these would be of a humanitarian nature and outside the sphere of conventional architectural practice. Instances would be migrant settlements, Favelas or responses to disasters. Also promote interdisciplinary thinking to help new generations of architects, designers and artists to anticipate the future through an educational programme of think tanks, symposia, films and publications and networks with selected universities, research institutions.

P100

P100

P36

©Cindy Palmano

P72
Zaha Hadid Architects
Zaha Hadid was born in Bagdad, Iraq in 1950. She graduated from the Architectural Association School of Architecture in 1972. She founded Zaha Hadid Architects in 1979. She is an architect who consistently pushes the boundaries of architecture and urban design. She completed her first building, the Vitra Fire Station, Germany in 1993. Her interest lies in the rigorous interface between architecture, urbanism, landscape and geology as her practice integrates natural topography and human-made systems, leading to innovation with new technologies. She taught Architectural Design at Yale University as a visiting professor and at the University of Applied Art in Vienna.

P52
Herzog & de Meuron
Was established in Basel, 1978. Has been operated by senior partners: Christine Binswanger, Ascan Mergenthaler and Stefan Marbach, with founding partners Pierre de Meuron and Jacques Herzog. An international team of about 40 Associates and 380 collaborators is working on projects across Europe, the Americas and Asia. The firm's main office is in Basel with additional offices in Hamburg, London, New York City, and Hong Kong. The practice has been awarded numerous prizes including the Pritzker Architecture Prize (USA) in 2001, the RIBA Royal Gold Medal (UK) and the Praemium Imperiale (Japan), both in 2007. In 2014, awarded the Mies Crown Hall Americas Prize (MCHAP).

P100
John Pawson [p.202, left-top]
Has spent more than 30 years making rigorously simple architecture, based on the qualities of proportion, light and materials. Key projects include Calvin Klein's first flagship store in New York, the award-winning Sackler Crossing in Kew's Royal Botanic Gardens and the new Cistercian monastery of Our Lady of Novy Dvur in Bohemia.

P36
The Chinese University of Hong Kong, School of Architecture
Kristof Crolla [p.202, left-bottom] graduated Magna Cum Laude as a civil architectural engineer at Ghent University in 2003. He practiced in the Belgium at Bureau Buildings & Techniques and designed his first built project, House for an Artist. Received his M. Arch from AA school, London. Worked for several years as a Lead Architect for Zaha Hadid Architects. Is currently a licensed architect who combines his position as an assistant professor in Computational Design at the Chinese University of Hong Kong with his architectural practice. The recent project, ZCB Bamboo Pavilion, won Small Project of the Year 2016 award at the World Architecture Festival.

Silvio Carta
Is an architect and researcher based in London. Received Ph.D. from University of Cagliari, Italy in 2010. His main fields of interest is architectural design and design theory. His studies have focused on the understanding of the contemporary architecture and the analysis of the design process. He taught at the University of Cagliari, Willem de Kooning Academy of Rotterdam and Delft University of Technology. He is now a senior lecturer at the University of Hertfordshire. Since 2008 he is an editor-at-large for C3, Korea and his articles have also appeared in A10, Mark, Frame and so on.

Studio Mumbai
Bijoy Jain was born in Mumbai, India in 1965 and received his M. Arch from Washington University in 1990. Worked in Richard Meier office at LA and London between 1989 and 1995. Returned to India in 1995 and founded Studio Mumbai. The work has been presented at the XII Venice Biennale and the Victoria & Albert Museum. Received Global Awards for Sustainable Architecture from the Institut Francais D'Architecture and Design for Asia Award from the Hong Kong Design Center in 2009. Has taught in Yale University, Architecture Academy of Mendrisio and so on.

Gihan Karunaratne
Is a British architect and a graduate of Royal College of Arts and Bartlett School of Architecture. Has taught and lectured in architecture and urban design in UK, Sri Lanka and China. Writes and researches extensively on art, architecture and urban design. Currently is the director of architecture for Colombo Art Biennale 2016. Has exhibited in Colombo Art Biennale in 2014, Rotterdam Architecture Biennale in 2009 and the Royal Academy Summer Exhibition. Is a recipient of The Bovis and Architect Journal award for architecture and was made a Fellow of Royal Society of Arts (RSA) for Architecture, Design and Education in 2012.

Herbert Wright
Graduated in Physics and Astrophysics from London University, and has worked in software publishing, press analysis and arts administration. He writes about architecture, urbanism and art for magazines and newspapers across Europe, and is a contributing editor of UK architecture magazine Blueprint. Launched and curated Lisbon's first public architectural event "Open House" in 2012. His first book London High (2006) was about London high-rise, and later book projects include collaborations with Dutch architects Mecanoo and Expo 2015 Gold Prize designer Wolfgang Buttress. He delivers occasional talks, tours and discussions.

Heidi Saarinen
Is a London based designer, lecturer at Coventry University and also an artist with current research focused on space and place. Is interested in the peripheral space, in-between and the interaction and collision between architecture, spaces, city, performance and the body. Is currently working on a series of interdisciplinary projects linking architecture, heritage, film and choreography in the urban environment. Is part of several community and creative groups in London and the UK where she engages in events and projects highlighting awareness of community and architectural conservation in the built environment.

© 2017大连理工大学出版社

版权所有·侵权必究

图书在版编目(CIP)数据

跨越时光的屋顶：英汉对照 / 英国扎哈·哈迪德建筑师事务所等编；蒋丽，贾子光译. — 大连：大连理工大学出版社，2017.11
 (建筑立场系列丛书)
 ISBN 978-7-5685-1129-2

Ⅰ. ①跨… Ⅱ. ①英… ②蒋… ③贾… Ⅲ. ①建筑设计－英、汉 Ⅳ. ①TU2

中国版本图书馆CIP数据核字(2017)第283910号

出版发行：大连理工大学出版社
　　　　　（地址：大连市软件园路80号　邮编：116023）
印　　刷：上海锦良印刷厂
幅面尺寸：225mm×300mm
印　　张：12.75
出版时间：2017年11月第1版
印刷时间：2017年11月第1次印刷
出 版 人：金英伟
统　　筹：房　磊
责任编辑：张昕焱
封面设计：王志峰
责任校对：杨　丹
书　　号：978-7-5685-1129-2
定　　价：258.00元

发　行：0411-84708842
传　真：0411-84701466
E-mail：12282980@qq.com
URL：http://dutp.dlut.edu.cn

本书如有印装质量问题，请与我社发行部联系更换。